Legume Crops - Prospects, Production and Uses

Edited by Mirza Hasanuzzaman

Published in London, United Kingdom

IntechOpen

Supporting open minds since 2005

Legume Crops - Prospects, Production and Uses
http://dx.doi.org/10.5772/intechopen.90304
Edited by Mirza Hasanuzzaman

Contributors
Kuldeep Tripathi, Padmavati Gore, Mamta Singh, Gayacharan C., Reena Mehra, Ravi Kishore Pamarthi, kangfu Yu, Eloisa Pajuelo, Salvadora Navarro-Torre, Khouloud Bessadok, Noris J. Flores-Duarte, Ignacio D. Rodríguez-Llorente, Miguel A. Caviedes, Zachary Shea, William Singer, Bo Zhang, Fatima Serebryanaya, Djavgarat Imachueva, Anita Mann, Ashwani Kumar, Satish Kumar Sanwal, Parbodh Chander Sharma, Suresh S. S Suryawanshi, Prajakta P. P. Kamble, Viawas A. Bapat, Jyoti P. Jadhav, Yuvaraj Muthuraman, M.Pandiyan Muthaiyan, Gayathri Pandurangan, Amanpreet Singh, Harmandeep Singh Chahal, Souframanien Jegadeesan

First published in London, United Kingdom, 2020 by IntechOpen
IntechOpen is the global imprint of INTECHOPEN LIMITED, registered in England and Wales, registration number: 11086078, 5 Princes Gate Court, London, SW7 2QJ, United Kingdom
Printed in Croatia

British Library Cataloguing-in-Publication Data
A catalogue record for this book is available from the British Library

Additional hard and PDF copies can be obtained from orders@intechopen.com

Legume Crops - Prospects, Production and Uses
Edited by Mirza Hasanuzzaman
p. cm.
Print ISBN 978-1-83968-273-5
Online ISBN 978-1-83968-274-2
eBook (PDF) ISBN 978-1-83968-275-9

For EU product safety concerns:
IN TECH d.o.o., Prolaz Marije Krucifikse Kozulić 3, 51000 Rijeka, Croatia, info@intechopen.com or visit our website at intechopen.com.

Meet the editor

Dr. Mirza Hasanuzzaman is a Professor of Agronomy at Sher-e-Bangla Agricultural University, Dhaka, Bangladesh. He received his PhD on 'Plant Stress Physiology and Antioxidant Metabolism' from the United Graduate School of Agricultural Sciences, Ehime University, Japan with a Japanese Government (MEXT) Scholarship. Later, he completed his postdoctoral research at the Center of Molecular Biosciences (COMB), University of the Ryukyus, Okinawa, Japan with a 'Japan Society for the Promotion of Science (JSPS)' postdoctoral fellowship. Subsequently, as an Adjunct Senior Researcher he joined the University of Tasmania with an Australian Government's Endeavour Research Fellowship. Prof. Hasanuzzaman has been devoting himself to research in the field of crop science, especially focused on environmental stress physiology since 2004. Prof. Hasanuzzaman has published over 100 articles in peer-reviewed journals and books. He has edited 13 books and written 45 book chapters on important aspects of plant physiology, plant stress responses, and environmental problems in relation to plant species. These books have been published by internationally renowned publishers. Prof. Hasanuzzaman is a research supervisor of undergraduate and graduate students and has supervised 20 M.S. students so far. He is editor and reviewer of more than 50 peer-reviewed international journals and the recipient of Publons Global Peer Review Award 2017, 2018 and 2019. He has been honored by different authorities due to his outstanding performance in different fields like research and education. He received the World Academy of Science (TWAS) Young Scientist Award in 2014. He has attended and presented 40 papers and posters in national and international conferences in different countries (USA, UK, Germany, Australia, Japan, Austria, Sweden, Russia, etc.).

Contents

Preface **XI**

Chapter 1 **1**
Legume Genetic Resources: Status and Opportunities for Sustainability
by Kuldeep Tripathi, Padmavathi G. Gore, Mamta Singh, Ravi K. Pamarthi, Reena Mehra and Gayacharan C

Chapter 2 **17**
Role of Legumes in Improving Soil Fertility Status
by Muthuraman Yuvaraj, Muthaiyan Pandiyan and Pandurangan Gayathri

Chapter 3 **29**
Soybean Production, Versatility, and Improvement
by Zachary Shea, William M. Singer and Bo Zhang

Chapter 4 **51**
Food Grade Soybean Breeding, Current Status and Future Directions
by Souframanien Jegadeesan and Kangfu Yu

Chapter 5 **79**
Organic Grain Legumes in India: Potential Production Strategies, Perspective, and Relevance
by Amanpreet Singh and Harmandeep Singh Chahal

Chapter 6 **97**
Sustainable Production of Pulses under Saline Lands in India
by Anita Mann, Ashwani Kumar, Satish Kumar Sanwal and Parbodh Chander Sharma

Chapter 7 **115**
Helping Legumes under Stress Situations: Inoculation with Beneficial Microorganisms
by Salvadora Navarro-Torre, Khouloud Bessadok, Noris J. Flores-Duarte, Ignacio D. Rodríguez-Llorente, Miguel A. Caviedes and Eloísa Pajuelo

Chapter 8 135
Bioactive Components of Magical Velvet Beans
by Suresh S. Suryawanshi, Prajakta P. Kamble, Vishwas A. Bapat and Jyoti P. Jadhav

Chapter 9 147
Hedysarum Species from Caucasus
by Serebryanaya Fatima and Imachueva Djavgarat

Preface

Legumes are plants from the Fabaceae family. Fabaceae is the third largest family among the angiosperms and they have vast economic importance. These plants provide food, fodder, fuel, medicine, and many other uses. Legume crops are a low-cost protein source and sometimes called the 'poor man's protein'. Legumes can fix nitrogen, and hence the requirement of exogenous nitrogen is low. Therefore, they are used as a soil enhancer, and some of the legume species are cultivated as green manure crops. In a sustainable agricultural system, legume crops are one of the essential components. However, improving the productivity of legume crops and improving their tolerance to adverse environments are important tasks for plant biologists.

This book presents nine chapters dealing with the importance, production and various uses of legumes. In the first chapter, the status of legume crops in the world and their prospects for sustainable agriculture are discussed. In the second chapter, many aspects of soil fertility improvement by legumes are discussed. The authors discuss how legumes play a role in maintaining soil health and nutrient status. In the third and fourth chapter, various aspects of soybean production and breeding, as well as their diverse use as food, are discussed. The fifth chapter presents some important aspects of organic grain legumes in India. In chapter 6, the approaches for soybean production under salinity are described. Salinity is one of the most devastating abiotic stressors for soybean. In this chapter, the authors show the plant responses and tolerance to salinity. In the next chapter, the authors show how beneficial microorganisms could confer stress tolerance in legumes. In chapter 8, the authors discuss the bioactive compounds from velvet beans. Chapter 9 presents a fascinating legume species, Hedysarum - their biology and production in the Caucasus.

I want to give special thanks to the authors for their outstanding and timely work in producing such excellent chapters. I am very thankful to Dajana Pemac and Lada Bozic, Author Service Managers at IntechOpen, for their prompt responses during the acquisition. I believe that this book will be useful for undergraduate and graduate students, teachers, and researchers, particularly from the field of Crop Science, Soil Science, Plant Breeding and Agronomy.

Mirza Hasanuzzaman
Sher-e-Bangla Agricultural University,
Dhaka, Bangladesh

Chapter 1

Legume Genetic Resources: Status and Opportunities for Sustainability

Kuldeep Tripathi, Padmavathi G. Gore, Mamta Singh, Ravi K. Pamarthi, Reena Mehra and Gayacharan C

Abstract

Legumes are one of the most valuable gifts of nature to man, animal, and environment. These are sustainable, affordable, water-efficient, and low-carbon footprint crop. Globally, the share of grain legume accessions is 15% of 7.4 million accessions conserved in genebanks, of which more than half of germplasm in genebanks are without characterization and evaluation data which ultimately limit the utilization of germplasm in legume improvement programs. Characterization of all genebank accessions should be of utmost priority for enhancing the utilization. The development of core, mini-core, reference sets, and trait-specific germplasm has provided route to crop breeders for mining genebanks. Identification of new sources of variation became easy with these subsets, but the entire collection also needs to be evaluated for unique and rare traits. In crop species with narrow genetic base, utilization of crop wild relatives as well as new resources aids to widen the genetic base of legume cultigens.

Keywords: core collection, crop wild relatives, diversity, genetic resources and legumes

1. Introduction

One of the biggest challenges in the twenty-first century is to produce sufficient nutritious food in the face of climate change, population explosion, and rapid urbanization and to do so in an environmentally sustainable manner. The United Nations Sustainable Development Goal number 2 explicitly recognizes the pivotal role that genetic diversity plays for food security, nutrition, and sustainable agriculture. Legumes, together with cereals, played a prominent role to the development of modern agriculture. The legume family, Fabaceae, is the third largest family of flowering plants, with 946 genera and 24,505 species [1]. Few domesticated ones have incontestably proven to be of crucial nutritional value for both humans and animals due to their protein content, causing them to be recognized as the second most valuable plant source of nutrients [2]. Legumes are extensively distributed in diverse agroclimatic zones globally, from alpine and arctic regions to the equatorial tropics.

The peculiar characteristics of the family are taproot system; bipinnate leaves; flower with corolla, standard petal (1), wing petal (2), and keel (2) (**Figure 1**); and

Figure 1.
Floral dissection of papilionaceous *family (*Vigna vexillata*): (a) standard petal, (b) wing petal, and (c) keel petal.*

S. no.	Tribe	Genus	a/b/c	Primary center of diversity
1	Phaseoleae	*Vigna*	107/26/24.3	Africa, Asia
		Dolichos	67/3/4.47	Africa, East Asia
		Phaseolus	50/3/6	America
		Cajanus	32/16/50	South Africa, South East Asia, and East Africa
		Macrotyloma	25/4/6.25	Tropical Africa and Asia
2	Cicereae	*Cicer*	44/5/8.8	Central Asia, Southwest Asia and Mediterranean countries, and Himalayas
3	Fabeae	*Lathyrus*	160/8/20	Europe, Asia and North America
		Vicia	160/15/10.6	Europe, Asia, and North America
		Lens	5/0/nil	Mediterranean countries and Central Asia
		Pisum	3/0/nil	Southern Europe, Mediterranean region, and West Asia

a, number of species present globally; b, number of species present in India; c, percentage of species found in India.
Source: [15].

Table 1.
List of important tribes with number of species.

root nodules that facilitate nitrogen fixation in the soil. The family is divided into three subfamilies, namely, Caesalpinoideae, Mimosoideae, and Papilionoideae [3]. Among them, subfamily Papilionoideae is economically very important as it contains most of the commercial leguminous crops. Among pulse crops, *Lathyrus* and *Vicia* are the largest genus comprising 160 species, which are naturally distributed

Sl. no.	Botanical name	Pulses	Oilseeds	Vegetable	Seed spice	Root/ tuber	Forage
1	*Arachis hypogaea* L. (peanut)		√				√
2	*Cajanus cajan* (L.) Millsp. (pigeonpea)	√		√			√
3	*Canavalia ensiformis* (L.) DC. (jack bean)	√					√
4	*Cicer arietinum* L. (chickpea)	√					
5	*Codariocalyx motorius* (Houtt.) H. Ohashi (telegraph plant)						√
6	*Cyamopsis tetragonoloba* (L.) Taub. (cluster bean)	√		√			√
7	*Flemingia procumbens* Roxb. (Sohphlang)					√	
8	*Glycine max* (L.) Merr. (soybean)	√	√				√
9	*Lablab purpureus* (L.) Sweet (hyacinth bean)	√		√			√
10	*Lathyrus sativus* L. (grass pea)	√		√			√
11	*Lens culinaris* Medik. (lentil)	√					
12	*Lupinus mutabilis* Sweet (tarwi)	√	√				√
13	*Macrotyloma uniflorum* (Lam.) Verdc. (horse gram)	√					√
14	*Medicago sativa* L. (lucerne)						√
15	*Melilotus indicus* (L.) All. (sweet clover)						√
16	*Periandra mediterranea* (Vell.) Taub. (sweet tuber)					√	
17	*Phaseolus lunatus* L. (lima bean)	√					
18	*Phaseolus vulgaris* L. (common bean)	√		√			√
19	*Pisum sativum* L. (garden pea)	√		√			√
20	*Psophocarpus tetragonolobus* (L.) DC. (winged bean)	√		√		√	√
21	*Psoralea esculenta* Pursh (bread root)					√	
22	*Pueraria tuberosa* (Willd.) DC. (Indian Kudzu)					√	

Sl. no.	Botanical name	Pulses	Oilseeds	Vegetable	Seed spice	Root/ tuber	Forage
23	*Trifolium alexandrinum* L. (Egyptian clover)						√
24	*Trifolium resupinatum* L. (Persian clover)						√
25	*Trigonella foenum-graecum* L. (fenugreek)			√	√		√
26	*Vicia faba* L. (faba bean)	√		√			√
27	*Vigna aconitifolia* (Jacq.) Marechal (moth bean)	√					
28	*Vigna angularis* (Willd.) Ohwi & H. Ohashi (adzuki bean)	√		√			
29	*Vigna mungo* (L.) Hepper (black gram)	√					
30	*Vigna radiata* (L.) R. Wilczek (green gram)	√					
31	*Vigna subterranea* (L.) Verdc. (bambara groundnut)	√				√	
32	*Vigna umbellata* (Thunb.) Ohwi & H. Ohashi (rice bean)	√		√			
33	*Vigna unguiculata* (L.) Walp. (cowpea)	√		√			√
34	*Vigna vexillata* (L.) A. Rich. (zombi pea)	√		√		√	√

Table 2.
List of important legume crops having diversified importance.

across temperate region of Europe, Asia, and America. *Vigna* has six subgenera. Ceratotropis is mostly referred to as Asiatic *Vigna*. Primary centers of diversity of major pulse genera are presented in **Table 1**.

Legumes play a major role in fulfilling human, animal food, and nutritional needs. The major grain legumes include dry beans, chickpea, cowpea, faba bean, lentil, field pea and pigeonpea, green gram, and black gram. Soybean and peanut are predominantly oil-producing legume. Beans, yard long bean, and garden pea where immature seeds and pods are eaten are vegetable types of legumes. Lucerne, berseem, grass pea, and cowpea are the forage legumes, while tuber legume includes zombi pea, winged bean, Sohphlang, etc. *Abrus precatorius* is having poisonous seeds which contain toxin *abrin.* In addition to these primary grain legumes, several underutilized potential legumes like cluster bean, horse gram, moth bean, and pillipsera, primarily grown in the Indian subcontinent, China, and South East Asia, are also equally important for ensuring food and nutritional security.

Legumes are the reservoir of the protein, carbohydrate, fiber, and minerals. In addition to that, low glycemic index (GI) makes them super food that provides long-term health benefits. The isoflavone content of legumes plays a role in plant defense [4] and root nodulation and is also beneficial in human health. In addition to the human health, legumes also serve as fodder for livestock. The best advantage of legumes is nitrogen fixation, through which they enrich soil fertility and improve soil texture for other crops, and, hence, plays an important role in cropping system [5]. Legumes with their economic use are presented as follows (**Table 2**).

2. Conservation

Despite their significant contribution in global food and nutrition security, it has also been reported that their production rate becomes static mainly due to biotic factors like viruses, insects, parasitic weeds, nematodes, fungi, bacteria, and abiotic factors, viz., extreme temperatures, drought, flood, mineral imbalance, etc. [6, 7]. Therefore, there is a need to explore sustainable alternative strategies to improve

Crop	USDA	NBPGR**	VIR	CGIAR institutes			SGSV*
				ICARDA	ICRISAT	IITA	
Chickpea	7000	14,626		15,246	20,764		47
Pigeonpea	4806	11,321			13,783		13,177
Green gram	3931	3935	856				8620
Pea	6161	4436		6113			11,929
Cowpea	1287	3671	1493			15,933	19,355
French bean	17,223	3919					42,314
Cluster bean		4051					61
Horse gram		2557					47
Rice bean		2032					
Zombi pea	32		1				460
Lucerne	1748	230					
Black gram	304	2200	220				363
Soybean	33,208	4779				4575	28,034
Grass pea		2613		4220			2537
Clovers		606		5152			—
Lentil		2515		12,477			13,591
Faba bean		866		10,029			4154
Groundnut		13,406			15,622		14,583
Bambara groundnut		2				1890	1528
Wild *Vigna*		565				1878	—
Adzuki bean		187					1562

**Ref. [16].*
***Ref. [17].*

Table 3.
Legume germplasm holdings in major genebanks.

and diversify their production. For that, the use of diverse legume genetic resources in crop improvement is one of the most sustainable strategies and ways to conserve valuable genetic resources for the future. Crop improvement programs are always depending upon genetic diversity available in genebank. Globally, genebanks hold ~1 million accessions of leguminous crop. Legume germplasm conserved in major genebank in the world has been presented in **Table 3**.

A large number of genetic resources are conserved ex situ in genebanks; a considerable amount of diversity remains untapped in the nature. Hence it became a priority to collect maximum amount of diverse germplasm before it lasts forever. Crop wild relatives (CWR) are reservoir of genes for breeding [8–10]. To explore the potential of CWRs in today's changing climate, collection and conservation become of utmost priority.

3. Utilization

For sustainable growth in agriculture production "Conservation through use" approach is the only way. Storing the genetic resources will not solve the purpose until it is utilized. In genebanks, genetic integrity is maintained over the periods with the aim to utilize this variability in the future and bring them to the mainstream breeding programs. More than 80% of genetic resources conserved in genebanks are without characterization and evaluation data. Huge collection size with large duplicates or triplicates is again a big constraint for systematic characterization and evaluation in multi-environment experiments. To tackle this situation, the concept of core collection [11] and mini-core collection [12] are considered as the best solution for characterization of samples that represent most of the variability of the germplasm collections. Core collection represents maximum genetic diversity with minimum repetitiveness of germplasm; hence, the size of germplasm became manageable without affecting the extent of genetic diversity of the germplasm (**Figure 2**).

A general procedure for the selection of a core collection can be divided into five steps, which are described in the following sections:

i. Identify the material (collection) that will be represented.

ii. Decide on the size of the core collection.

iii. Divide the set of material used into distinct groups.

iv. Decide on the number of entries per group.

v. Choose the entries from each group that will be included in the core.

Conventional core and mini-core collections have been developed in many legume crops. **Table 4** represents the core and mini-core developed in legumes.

Trait-specific reference set is also developed by various genebanks which offers huge opportunities to identify novel sources of variation for use in breeding program. Discovery of new traits is also possible during large-scale characterization program which resulted into unique genotypes for its further exploitation in breeding programs. For example, unique seed morphotype with extended funiculus was found during lentil characterization of 2600 accessions of lentil, and this trait is associated with fast water uptake [13].

Figure 2.
Field view of lentil characterization program at ICAR-NBPGR, India.

Sl. no.	Crop	Core/mini-core	References
1	Soybean	Core and mini-core	[18–23]
2	Peanut	Core and mini-core	[24–28]
3	Chickpea	Core and mini-core	[12, 29, 30]
4	Pigeonpea	Core and mini-core	[30, 31]
5	Lentil	Core	[30, 32]
6	Mungbean	Core and mini-core	[33, 34]
7	Adzuki bean	Core	[35]
8	Common bean	Core	[36–42]
9	Cowpea	Core	[43]
10	Moth bean	Core	[44]
11	Pea	Core	[45]
12	Hyacinth bean	Core	[46]
13	*Medicago* spp.	Core	[47–49]

Table 4.
List of cores and mini-cores developed in legume crops.

Crop wild relatives (CWR) are wild plant species genetically more or less closely related to a particular crop, but unlike the crop species has not been domesticated and remain untouched by humans. Being progenitors of crop, they contain enormous genetic variation, which are readily available to plant breeders to use in crop improvement programs and to meet the challenge of global food security along with enhancing agricultural production and sustainability in the context of a rapidly growing world population and accelerated climate change. CWRs can be categorized based on the genecology that explains the extent to which CWRs can exchange genes with the crop. The Taxon Group (TG) concept is as follows: TG1a comprises crop species; TG1b, the taxa within the same species as crop; TG2, taxa in the same series or section as crop; TG3, taxa within the same subgenus as crop; TG4, taxa

Crop	Gene pool			References
	GP1	GP2	GP3	
Chickpea	*C. reticulatum*	*C. echinospermum*	*C. judaicum*, *C. pinnatifidum*, *C. bijugum*, *C. cumeatum*, *C. chorassanicum*, and *C. yamashitae*	[50]
Cluster bean	*C. senegalensis*	—	—	[50]
Lentil	*L. culinaris*, *L. orientalis*, *L. tomentosus*	*L. lamottei*, *L. odemensis*	*L. ervoides*	[51]
Green gram	*V. radiata* var. *radiata*, *V. radiata* var. *sublobata*, *V. radiata* var. *setulosa*	*V. mungo* var. *mungO*, *V. mungo* var. *silvestris*, *V. aconitifolia*, *V. trilobata*	*V. angularis*, *V dalzelliana*, *V. glabrescens*, *V. grandis*, *V. umbellata*, *V. vexillata*	[14, 52–55]
Black gram	*V. mungo* var. *mungo*, *V. mungo* var. *silvestris*	*V. radiata* var. *radiata*, *V. radiata* var. *sublobata*, *V. radiata* var. *setulosa*, *V. aconitifolia*, *V. trilobata*	*V. angularis*, *V. dalzelliana*, *V. glabrescens*, *V. grandis*, *V. umbellata*, *V. vexillata*	[14, 52, 53]
Cowpea	*V. unguiculata* var. *unguiculata* (L.) Walp (cv. *unguiculata*, cv. *biflora*, cv. *sesquipedalis*, cv. *melanophthalmus*, cv. *textilis*), *V. unguiculata* var. *spontanea* (Schweinf.), *V. unguiculata* subsp. *alba*, *V. unguiculata* subsp. *dekinditiana* (Harms.), *V. unguiculata* subsp. *pubescence*, *V. unguiculata* subsp. *stenophylla*, *V. unguiculata* subsp. *tenuis*	*V. unguiculata* subsp. *aduensis*, *V. unguiculata* subsp. *baoulensis*, *V. unguiculata* subsp. *burundiensis*, *V. unguiculata* subsp. *letouzeyi*, *V. unguiculata* subsp. *pawekiae*		[55]
Faba bean	—	*V. narbonensis*, *V. hyaeniscyamus*, *V. galilaea*, *V. johannis*, *V. bithynic*		[56]
Pigeonpea	*C. cajanifolius*	*C. lineatus*, *C. sericeus*, *C. scarabaeoides*, *C. albicans*, *C. trinervius*, *C. reticulatus*, *C. confertiflorus*, *C. latisepalous*	*C. platycarpus*, *C. lanceolatus*, *C. acutifolius*	[57]
Adzuki bean	*V. angularis* var. *nipponensis* and wild types of *V. umbellata*	*V. dalzelliana*, *V. glabrescence*, *V. minima*	*V. aconitifolia*, *V. mungo*, *V. radiata*, *V. trilobata*, *V. grandis*	[14]
Rice bean	*V. angularis*	*V. dalzelliana*, *V. glabrescence*, *V. minima*	*V. aconitifolia*, *V. mungo*, *V. radiata*, *V. trilobata*, *V. grandis*	[14]

Table 5.
Gene pool of major legumes.

Crop	Wild relative	Trait	Reference
Chickpea	*C. microphyllum*	Resistant to legume pod borer (*Helicoverpa armigera)*	[57]
	C. echinospermum, *C. pinnatifidum*, *C. bijugum*, *C. judaicum*, and *C. montbretii*	*Ascochyta* blight resistance	[58, 59]
	C. bijugum, *C. judaicum*, *C. pinnatifidum*, *C. reticulatum*, *C. echinospermum*, and *C. cuneatum*	*Fusarium* wilt resistance	[59, 60]
	C. judaicum, *C. pinnatifidum*	Gray mold resistance	[60]
	C. echinospermum	*Phytophthora* root rot	[61]
	C. pinnatifidum, *C. bijugum*, *and C. reticulatum*	Cyst nematode	[59]
Green gram	*V. aconitifolia*, *V. glabrascence*, *V. sublobata V. umbellate*	*Cercospora* leaf spot disease resistant	[62]
	V. radiata var. sublobata	Bruchid resistant	[62, 63]
	V. luteola, *V. trilobata*	Salt stress resistant	[64]
Black gram	*V. mungo* var. *silvestris*	Mungbean yellow mosaic virus (MYMV) resistance	[65]
	V. mungo var. *silvestris*	Bruchid resistance	[65]
Cowpea	*V. pubescence*, *V. vexillata*, *V. reticulata*, *V. oblongifolia*, *V. luteola*	Insects Resistance	[65]
	V. ambacensis, *V. davyi*, *V. glabrescens*, *V. marina*, *V. mungo*, *V. oblongifolia*, *V. parkeri*, *V. racemosa*, *V. reticulata*, *V. vexillata*, and *V. unguiculata* subsp. *dekindtiana*	Resistance to *Striga gesnerioides*	[66]
Pigeonpea	*C. scarabaeoides*, *C. sericeus*, *C. acutifolius*, *C. lineatus*, *C. albicans*	Pod fly (*Melanagromyza obtusa*)	[67]
	C. scarabaeoides, *C. albicans*	Pod wasp (*Tanaostigmodes cajaninae*)	[68]
	C. acutifolius	Sterility mosaic disease resistance, salt tolerance	[68]
	C. albicans	Pod borer resistance, sterility mosaic disease resistance, salt tolerance	[68]
	C. cajanifolius	Nuclear male sterility, cytoplasmic male sterility	[68]
	C. sericeus	Cytoplasmic male sterility, *Phytophthora* blight resistance	[68]
Pea	*P. fulvum*	Pea weevil, rust, powdery mildew and *Ascochyta* blight	[69]
	P. sativum subsp. *elatius*	Resistant to nematodes, weevil, broomrape, powdery mildew, *Fusarium* wilt, root rot, *Ascochyta* blight and white wilt	[69]

Table 6.
Wild genetic resources as trait donor in few pulse crops [70–77].

within the same genus as crop; and TG5, different genus to the crop [8]. CWRs have been categorized into three gene pools as primary gene pool (GP1) contains close relatives that readily intercross with the crop. Secondary gene pool (GP2) contains all the biological species that can be crossed with the crop but where hybrids are usually sterile. Tertiary gene pool (GP3) comprises those species that can be crossed with the crop with difficulty and where gene transfer is only possible with radical techniques. Another way is taxonomic which is based on taxonomic relationship of CWR with the crop [12]. Gene pools of some of the major legumes are represented (**Table 5**).

CWRs have provided vital genetic diversity for crop improvement since the twentieth century. They imparted resistance to numerous pests and diseases and tolerance to many abiotic stresses, viz., extreme temperatures, drought, and flood, and to improve nutrition, flavor, color, texture, and yield stability [13]. Almost all modern varieties of crops contain one or more genes derived from a CWR and contributed significantly to the agricultural and horticultural industries and to the world economy [14]. Furthermore, being components of natural ecosystems, they also play a role in functioning and maintaining the ecosystem services. However, many of CWRs remain unexplored. To explore the unexplored potential of CWRs, collection, conservation, characterization, and evaluation are the only powerful ways. Some examples of the use of CWRs in providing resistance to abiotic and biotic stress yield and quality improvement are listed in **Table 6**.

4. Conclusions

Genetic resources are recognized as the fourth essential input after water, soil, and light. It is the need of hour to utilize leguminous genetic resources to meet the global challenges like population explosion, malnutrition, and hunger. Legumes are dominated by rice-wheat system in the past with the major focus on cereal production worldwide. This has led to a substantial decrease in research on legumes. In view of the current climate change scenario and environmental disturbances, research on landraces and CWR is strengthened in effective and efficient manner. Legumes have the potential to contribute significantly to eco-friendly agricultural land use and sustainable forage production in the tropics. Earlier, genebanks were focusing on the conservation of genetic resources, but emphasis has been shifted to conservation through utilization. Overall, legume genetic resources provide the new potential climate smart crops, viz., zombi pea, winged bean, and grass pea, and also new alleles which help in developing biotic and abiotic stress-tolerant varieties. It is much needed to identify the bottleneck for the utilization of the valuable germplasm of the legumes. With the advancement of the modern molecular technologies, the trait discovery and the markers linked to the traits need to be explored so that a large-scale screening would become possible and eventually help to reveal the real genetic potential of the germplasm conserved in the gene banks.

Acknowledgements

We greatly thank Director of ICAR-NBPGR for his guidance in the management of legume genetic resources at ICAR-NBPGR, New Delhi, India.

Conflict of interest

The authors declare no conflict of interest.

Author details

Kuldeep Tripathi[1*], Padmavathi G. Gore[1], Mamta Singh[1], Ravi K. Pamarthi[1], Reena Mehra[2] and Gayacharan C[1]

1 ICAR-National Bureau of Plant Genetic Resources, New Delhi, India

2 International Centre for Agricultural Research in Dry Areas, Food Legumes Research Platform, Sehore, India

*Address all correspondence to: kuldeep.tripathi@icar.gov.in

References

[1] The Plant List. Available from: http://www.plantlist.org/1.1/browse/A/Leguminosae/ [Accessed: 24 January 2020]

[2] Bhat R, Karim AA. Exploring the nutritional potential of wild and underutilized legumes. Comprehensive Reviews in Food Science and Food Safety. 2009;**8**:305-331

[3] Kumar S, Sane PV. Legumes of South Asia, A Check-List. Kew: Royal Botanic Gardens; 2003. p. 536

[4] Padmavati M, Reddy AR. Flavonoid biosynthetic pathway and cereal defence response: An emerging trend in crop biotechnology. Journal of Plant Biochemistry and Biotechnology. 1999;**8**:15-20

[5] Graham PH, Vance CP. Legumes: Importance and constraints to greater use. Plant Physiology. 2003;**131**(3):872-877

[6] Dwivedi SL, Upadhyaya HD, Hegde DM. Development of core collection using geographic information and morphological descriptors in safflower (*Carthamus tinctorius* L.) Germplasm. Genetic Resources and Crop Evolution. 2005;**52**:821-830

[7] Ojiewo C, Monyo E, Desmae H, Boukar O, Mukankusi-Mugisha C, Thudi M, et al. Genomics, genetics and breeding of tropical legumes for better livelihoods of smallholder farmers. Plant Breeding. 2019;**138**:487-499

[8] Maxted N, Ford-Lloyd BV, Jury SL, Kell SP, Scholten MA. Towards a definition of a crop wild relative. Biodiversity and Conservation. 2006;**15**(8):2673-2685

[9] Gore PG, Rana MK, Tripathi K, Singh M, Bisht IS, Bhalla S. Genetic diversity in wild *Lens* spp. using inter simple sequence repeat (ISSR) marker. Legume Research. 2015;**38**(5):575-578

[10] Gore PG, Tripathi K, Pratap A, Bhat KV, Umdale SD, Gupta V, et al. Delineating taxonomic identity of two closely related *Vigna* species of section *Aconitifoliae*: *V. trilobata* (L.) Verdc. and *V. stipulacea* (Lam.) Kuntz in India. Genetic Resources and Crop Evolution. 2019;**66**:1155-1165

[11] Upadhyaya HD, Oritz RA. Mini core subset for capturing diversity and promoting utilization of chickpea genetic resources in crop improvement. Theoretical and Applied Genetics. 2001;**102**(8):1292-1298

[12] Harlan JR, de Wet JM. Toward a rational classification of cultivated plants. Taxon. 1971;**20**(4):509-517

[13] Hajjar R, Hodgkin T. The use of wild relatives in crop improvement: A survey of developments over the last 20 years. Euphytica. 2007;**156**(1):1-13

[14] Dana S, Karmakar PG. Species relation in Vigna subgenus *Ceratotropis* and its implications in breeding. Plant Breeding Reviews. 1990;**8**:19-42

[15] Pradheep K, Bhandari DC, Bansal KC. Wild Relatives of Cultivated Plants in India. New Delhi: Indian Council of Agricultural Research; 2014. pp. 728

[16] SGSV Database. Available from: https://www.nordgen.org/sgsv/index [Accessed: 26 January 2020]

[17] ICAR-NBPGR. Available from: http://genebank.nbpgr.ernet.in/CropgroupCrops.aspx [Accessed: 26 January 2020]

[18] Oliveira MF, Nelso RL, Geraldi IO, Cruz CD, de Toledo JFF. Establishing a soybean germplasm core collection.

Field Crops Research. 2010;**119**(2-3): 277-289

[19] Qiu L, Cao Y, Chang R, Zhou X, Wang G, Sun J, et al. Establishment of Chinese soybean (*Glycine max*) core collection. Sampling strategy. Zhongguonongyekexue. 2002;**36**(12): 1442-1449

[20] Li JQ, Ying HL, Rong XG, Zhang XL, Li XW, Ru ZC. Establishment, representative testing and research progress of soybean core collection and mini core collection. Acta Agronomica Sinica. 2009;**35**(4):571-579

[21] Wang L, Li Y, Li W, Zhu L, Guan Y, Ning X, et al. Establishment of a core collection of Changjiang spring sowing soybean. Chinese Biodiversity. 2003;**12**(6):578-585

[22] Limei Z, Yingshan D, Bao L, Shui H, Kejing W, Xianghua L. Establishment of a core collection for the Chinese annual wild soybean (*Glycine soja*). Chinese Science Bulletin. 2005;**50**(10):989-996

[23] Brown AHD, Grace JP, Speer SS. Designation of a core collection of perennial glycine. Soybean Genetics Newsletter. 1987;**14**:59-70

[24] Holbrook CC, Anderson WF, Pittman RN. Selection of a core collection from the U.S. germplasm collection of peanut. Crop Science. 1993;**33**:859-861

[25] Upadhyaya HD, Bramel PJ, Ortiz R, Singh S. Developing a mini core of peanut for utilization of genetic resources. Crop Science. 2002;**42**(6):2150-2156

[26] Dwivedi SL, Puppala N, Upadhyaya HD, Manivannan N, Singh S. Developing a core collection of peanut specific to Valencia market type. Crop Science. 2008;**48**:625-632

[27] Upadhyaya HD, Ortiz R, Bramel PJ, Singh S. Development of a groundnut core collection using taxonomical, geographical and morphological descriptors. Genetic Resources and Crop Evolution. 2003;**50**(2):139-148

[28] Hui-Fang J, Xiao-Ping R, Bo-Shou L, Jia-Quan H, Yong L, Ben-Yin C, et al. Peanut core collection established in China and compared with ICRISAT mini core collection. Acta Agronomica Sinica. 2008;**34**(1):25-30

[29] Upadhyaya HD, Bramel PJ, Singh S. Development of a chickpea core subset using geographic distribution and quantitative traits. Crop Science. 2001;**41**(1):206-210

[30] Meena HP, Kumar J, Upadhyaya HD, Bharadwaj C, Chauhan SK, Verma AK, et al. Chickpea mini core germplasm collection as rich sources of diversity for crop improvement. Journal of SAT Agricultural Research. 2010;**8**:1-5

[31] Reddy LJ, Upadhyaya HD, Gowda CLL, Singh S. Development of core collection in pigeon pea [*Cajanus cajan* (l.) Millsp.] using geographical and qualitative morphological descriptors. Genetic Resources and Crop Evolution. 2005;**52**:1049-1056

[32] Tullu A, Kusmenoglu I, McPhee KE, Muehlbauer FJ. Characterization of core collection of lentil germplasm for phenology, morphology, seed and straw yields. Genetic Resources and Crop Evolution. 2001;**48**(2):143-152

[33] Bisht IS, Mahajan R, Patel RK. The use of characterisation data to establish the Indian mungbean core collection and assessment of genetic diversity. Genetic Resources and Crop Evolution. 1998;**45**(2):127-133

[34] Schafleitner R, Nair RM. The AVRDC—The world vegetable center mungbean (*Vigna radiata*) core and

mini core collections. BMC Genomics. 2015;**16**(1):1

[35] Wang SM, Tan FJ, Hu JP. Genetic diversity assessment of adzuki bean germplasm resources based on isozyme analysis. Scientia Agricultra Sinica. 2002;**35**:1311-1318

[36] Tohme J, Jones P, Beebe S, Iwanaga M. The combined use of agroecological and characterization data to establish the CIAT *Phaseolus vulgaris* core collection. In: Hodgkin T, Brown AHD, van Hintum TJL, Morales EAV, editors. Core Collections Genetic Resources. Chichester, UK: John Wiley & Sons; 1995. pp. 95-107

[37] Skroch PW, Nienhuis J, Beebe S, Tohme J, Pedraza F. Comparison of Mexican common bean (*Phaseolus vulgaris* L.) core and reserve germplasm collections. Crop Science. 1998;**38**:488-496

[38] Zeven A, Waninge J, Hintum TV, Singh SP. Phenotypic variation in a core collection of common bean (*Phaseolus vulgaris* L.) in the Netherlands. Euphytica. 1999;**109**:93-106

[39] Logozzo G, Donnoli R, Macaluso L, Papa R, Knupffer H, ZPL S. Analysis of the contribution of Mesoamerican and Andean gene pools to European common bean (*Phaseolus vulgaris* L.) germplasm and strategies to establish a core collection. Genetic Resources and Crop Evolution. 2007;**54**:1763-1779

[40] Perez-Vega E, Campa A, De La RL, Giraldez R, Ferreira JJ. Genetic diversity in a core collection established from the main bean genebank in Spain. Crop Science. 2009;**49**(4):1377-1386

[41] Blair MW, Díaz LM, Buendía HF, Duque MC. Genetic diversity, seed size associations and population structure of a core collection of common beans (*Phaseolus vulgaris* L.). Theoretical and Applied Genetics. 2009;**119**(6):955-972

[42] Rodino AP, Santalla M, De Ron AM, Singh SP. A core collection of common bean from the Iberian peninsula. Euphytica. 2003;**131**(2):165-175

[43] Mahalakshmi V, Ng Q, Lawson M, Ortiz R. Cowpea [*Vigna unguiculata* (L.) Walp.] core collection defined by geographical, agronomical and botanical descriptors. Plant Genetic Resources: Characterization and Utilization. 2007;**5**(03):113-119

[44] Meghwal R, Joshi U, Kumar S, Sharma R. Screening of moth bean (*Vigna aconitifolia*) core collection against yellow mosaic virus. Indian Journal of Agricultural Sciences. 2005;**85**(4):571-575

[45] Xu-Xiao Z, Jian-Ping G, Shu-min W, Qin-chang L, Robert RR, Rebecca F. Genetic diversity and core collection of alien *Pisum sativum* L. germplasm. Acta Agronomica Sinica. 2008;**34**(9):1518-1528

[46] Pengelly BC, Maass BL. *Lablab purpureus* (L.) sweet–diversity, potential use and determination of a core collection of this multi-purpose tropical legume. Genetic Resources and Crop Evolution. 2001;**48**(3):261-272

[47] Diwan N, McIntosh MS, Bauchan GR. Methods of developing a core collection of annual *Medicago* species. Theoretical and Applied Genetics. 1995;**90**(6):755-761

[48] Basigalup DH, Barnes DK, Stucker RE. Development of a core collection for perennial *Medicago* plant introductions. Crop Science. 1995;**35**(4):1163-1168

[49] Ellwood SR, D'souza NK, Kamphuis LG, Burgess TI, Nair RM, Oliver RP. SSR analysis of the *Medicago truncatula* SARDI core collection reveals

substantial diversity and unusual genotype dispersal throughout the Mediterranean basin. Theoretical and Applied Genetics. 2006;**112**(5):977-983

[50] Smartt J. Gene pools in grain legumes. Economic Botany, JSTOR. 1984;**38**(1):24-35

[51] Wong MML, Verma NG, Ramsay L, Yuan HY, Caron C, Diapari M, et al. Classification and characterization of species within the genus lens using genotyping-by-sequencing (GBS). PLoS One. 2015;**10**(3):e0122025. DOI: 10.1371/ journal.pone.0122025

[52] Smartt J. Evolution of grain legumes. III. Pulses in the genus *Vigna*. Experimental Agriculture. 1985;**21**:87

[53] Smartt J. Gene pools in Phaseolus and Vigna cultigens. Euphytica. 1981;**30**:445-459

[54] Kumar S, Gupta S, Chandra S, Singh BB. How wide is the genetic base of pulse crops? In: Ali M, Singh BB, Kumar S, Dhar V, editors. Pulses in New Perspective. Kanpur, India: Indian Society of Pulses Research and Development; 2004. pp. 211-221

[55] Kumar S, Imtiaz M, Gupta S, Pratap A. Distant hybridization and alien gene introgression. In: Pratap A, Kumar J, editors. Biology and Breeding of Food Legumes. Oxfordshire, UK: CABI; 2011. pp. 81-110

[56] Singh AK, Bhatt BP. Faba bean: Unique germplasm explored and identified. Hort Flora Research Spectrum. 2012;**3**:267-269

[57] Singh R, Sharma P, Varshney RK, Sharma SK, Singh NK. Chickpea improvement: Role of wild species and genetic markers. Biotechnology and Genetic Engineering Reviews. 2008;**25**:267-314

[58] Singh KB, Hawtin GC, Nene YL, Reddy MV. Resistance in chickpeas to *Ascochyta rabiei*. Plant Diseases. 1981;**65**:586-587

[59] Singh KB, Ocampo B, Robertson LD. Diversity for abiotic and biotic stress resistance in the wild annual *Cicer* species. Genetic Resources and Crop Evolution. 1998;**45**:9-17

[60] Nene YL, Haware MP. Screening chickpea for resistance to wilt. Plant Diseases. 1980;**64**:379-380

[61] Singh G, Kapoor S, Singh K. Screening chickpea for gray mold resistance. International Chickpea Newsletter. 1982;**7**:13-14

[62] Tomooka N, Lairungruang C, Nakeeraks P, Egawa Y, Thavarasook C. Development of bruchid resistant mungbean using wild mungbean germplasm in Thailand. Plant Breeding. 1992;**109**:60-66

[63] Watanasit A, Pichitporn S. Improvement of mungbean for resistance to bruchids. In: Srinives P, Kitbamroong C, Miyazaki S, editors. Mungbean Germplasm: Collection, Evaluation and Utilization for Breeding Program. Tsukuba, Japan: Japan International Research Center for Agricultural Sciences; 1996. pp. 67-71

[64] Nirmala S, Bhat KV, Sairam RK, Jaiwal PK. Identification of salt resistant wild relatives of mungbean (*Vigna radiata* (L.) Wilczek). Asian Journal of Plant Science and Research. 2013;**3**(5):41-49

[65] Souframanien J, Gopalakrishna T. ISSR and SCAR markers linked to the mungbean yellow mosaic virus (MYMV) resistance gene in black gram [*Vigna mungo* (L.) Hepper]. Plant Breeding. 2006;**125**:619-622

[66] Oyatomi C, Fatokun O, Boukar M, Abberton C. Screening wild vigna species and cowpea (*Vigna unguiculata*) landraces for sources of resistance

to *Striga gesnerioides*. In: Maxted N, Dulloo ME, Ford-Lloyd BV, editors. Enhancing Crop Genepool Use: Capturing Wild Relative and Landrace Diversity for Crop Improvement. University of Birmingham, UK: CAB International; 2016

[67] Sharma HC, Pampapathy G, Reddy LJ. Wild relatives of pigeonpea as a source of resistance to the pod fly (*Melanagromyza obtusa* Malloch) and pod wasp (*Tanaostigmodes cajaninae* La Salle). Genetic Resources and Crop Evolution. 2003;**50**(8):817-824

[68] Khoury CK, Castañeda-Alvarez NP, Achicanoy HA, Sosa CC, Bernau V, Kassa MT, et al. Crop wild relatives of pigeonpea [*Cajanus cajan* (L.) Millsp.]: Distributions, ex situ conservation status, and potential genetic resources for abiotic stress tolerance. Biological Conservation. 2015;**184**:259-270

[69] Kosterin OE. Prospects of the use of wild relatives for pea breeding. Russian Journal of Genetics and Applied Research. 2016;**6**:233-243

[70] Frankel OH, Brown AHD. Plant Genetic resources today; A critical appraisal. In: Holden JHW, Williams JT, editors. Crop Genetic Resources Conservation and Evaluation. London: George Allen and Unwin Ltd.; 1984. pp. 249-257

[71] Holbrook C, Dong W. Development and evaluation of a mini core collection for the U.S. peanut germplasm collection. Crop Science. 2005;**46**:1540-1544

[72] Upadhyaya HD, Reddy LJ, Gowda CLL, Reddy KN, Singh SS. Development of a mini core subset for enhanced and diversified utilization of pigeonpea germplasm resources. Crop Science. 2006;**46**:2127-2132

[73] Phillip EM, Jeff T, Melody M, Caleb W, Rian L, Sujan M. Population structure and genetic differentiation among the USDA common bean (*Phaseolus vulgaris* L.) core collection. Genetic Resources and Crop Evolution. 2012;**59**(4):499-515

[74] Tripathi K, Gore PG, Pandey A, Bhardwaj R, Singh N, Chawla G, et al. Seed morphology, quality traits and imbibition behaviour study of atypical lentil (*Lens culinaris* Medik.) from Rajasthan, India. Genetic Resources and Crop Evolution. 2019;**66**(3):697-706

[75] Maxted N, Kell SP. Establishment of a Global Network for the In Situ Conservation of Crop Wild Relatives: Status and Needs. Rome, Italy: FAO Commission on Genetic Resources for Food and Agriculture; 2009. p. 112

[76] Anderson SM, Vicente M. Gene Flow between Crops and Their Wild Relatives. Baltimore: John Hopkins University Press; 2010

[77] Available from: http://www.icrisat.org/what-we-do/crops/Pigeon Pea/Archives/uwsppi.htm [Accessed: 24 January 2020]

Chapter 2

Role of Legumes in Improving Soil Fertility Status

Muthuraman Yuvaraj, Muthaiyan Pandiyan and Pandurangan Gayathri

Abstract

Legume plants have a probably important role to play in growing indigenous nitrogen production besides meeting human demands for protein and energy. Some legumes have the capability to solubilize in any other case unavailable phosphate by excreting organic acids from their roots, in addition to improving soil fertility. Legumes also assist to restoration of soil natural matter and limit pest and disease issues when used in rotation with nonleguminous crops. Research has shown that the organic nitrogen fixation procedure is the most environment friendly way to grant the giant amounts of nitrogen wished through legumes to produce high-yielding crops with an excessive protein content. For the fixation technique to occur, legume vegetation must enter into a "symbiotic" or collectively beneficial partnership with sure microorganism known as rhizobia. Soon after legume seeds germinate, rhizobia current in the soil or delivered as seed inoculum invade the root hairs and go through an infection thread toward the root. The bacteria multiply rapidly in the root, causing the swelling of root cells to structure nodules.

Keywords: legume, soil fertility, greenhouse gas, crop rotation

1. Introduction

Global populace will hit 9.6 billion human beings with the aid of 2050 [1] and will face world challenges among which attaining meals security, reducing the risk of local weather exchange through lowering the net release of greenhouse gases into the ecosystem and assembly the increasing demand for energy are the most critical ones. In particular, the impact of climate trade and related biotic and abiotic stresses to which crop structures will be an increasing number of uncovered pose serious implications for global food production [2].

To meet these challenges, a policy framework needs to be developed in which the sustainability of production consumption patterns turns into central. In this context, meal legumes and legume-inclusive manufacturing systems can play essential roles by means of turning in more than one offerings in line with sustainability principles. Indeed, legumes play central roles [3]: (a) at food-system level, both for human and animal consumption, as a source of plant proteins and with an increasingly importance in enhancing human beings health [4]; (b) at production-system level, due to the capability to fix atmospheric nitrogen making them potentially notably appropriate for inclusion in low-input cropping systems, and due to their function in mitigating greenhouse gases emissions [5]; and (c) at cropping-system levels, as diversification vegetation in agroecosystems primarily based on few important species, breaking the cycles of pests and diseases and contributing to stability the deficit in plant protein manufacturing in many areas of the world.

Legumes have a probably substantial position to play in enhancing soil carbon sequestration. They can also have considerable additional advantages beyond their significance involving nitrogen fixation and excessive protein feeds. These consist of advantageous impacts on biodiversity and soil quality. There is a great need for a strong focus on creating the role of legumes and their contribution to each the sustainable intensification of manufacturing and the livelihoods of small holder farmers in many components of the world [6]. Apart from their makes use of as food and fodder they have a very necessary position in retaining soil fertility by fixing atmospheric nitrogen and enhancing soil structures and adding organic matters. Moreover, it is generally used as an intercrop and covers plants, and sometimes, it is cultivated as emergency vegetation due to its brief life cycle. Since it requires low fertilizer and other inputs this crop is relatively profitable in a most economical point of view. It also improves environmental quality by sequestrating carbon and mitigating other pollutants. Legumes are additionally a potential plant team in which some of the species having a capacity of remediating poisonous metals and organic pollutants [7].

2. Nitrogen fixation

Legume plant and seed tissue is distinctly high in protein. This can be without delay attributed to a legume's capability to supply most of its personal nitrogen wants with the assist of symbiotic Rhizobia microorganism residing in their roots. Inoculated with the applicable stress of Rhizobia bacteria, legumes can furnish up to 90% of their own nitrogen (N). Shortly after a legume seed germinates in the presence of Rhizobia microorganism in the soil, the bacteria penetrate the root hairs and cross into the root itself. The bacteria multiply, inflicting a swelling of the root to shape pale pink nodules. Nitrogen gasoline present in the soil air is then sure by the microorganism which feed on carbohydrates manufactured by the above-ground plant in the course of photosynthesis [8]. The bacteria produce ammonia (NH_3) from the hydrogen obtained from the plant's carbohydrates and nitrogen from the air. The ammonia then provides a supply of nitrogen for the plant to grow. This symbiotic relationship between bacteria and legume lets in them both to flourish and produce a high-protein seed or forage crop. Even although legumes can repair nitrogen from the atmosphere, they can take up large quantities of soil nitrogen if it is available. Nitrogen release from a legume crop occurs as the above-ground plant residues, roots and nodules step by step decompose. Soil microorganisms decompose the highly nitrogen-rich organic cloth and launch the nitrogen to the soil when they die. Usually about two-thirds of the nitrogen fixed through a legume crop becomes handy the subsequent growing season after a legume in a rotation [9].

3. Advantages of legumes in soil quality

Soil quality advantages of legumes include increasing soil natural matter, improving soil porosity, recycling nutrients, improving soil structure, decreasing soil pH, diversifying the microscopic lifestyles in the soil, and breaking disease build-up and weed problems of grass-type crops.

3.1 Soil natural rely

As stated previously, legumes are high in protein, and therefore, nitrogen rich. Because most crop residues incorporate a lot extra carbon than nitrogen, and microorganism in the soil need both, the nitrogen provided by legumes allows the decomposition of crop residues in the soil and their conversion to soil constructing natural matter.

3.2 Soil porosity

Several legumes have aggressive taproots reaching 6–8 feet deep and a half inch in diameter that open pathways deep into the soil. Nitrogen-rich legume residues inspire earthworms and the burrows they create. The root channels and earthworm burrows make bigger soil porosity, promotion air movement and water percolation deep into the soil.

3.3 Recycle vitamins

Because perennial and biennial legumes root deeply in the soil, they have the capability to recycle crop nutrients that are deep in the soil profile. This effects in a more environment friendly use of utilized fertilizer and prevents nutrients (particularly nitrate nitrogen) from being lost due to leaching under the root region of shallower-rooted crops in the rotation (**Figure 1**).

3.4 Improve soil structure

The improvements are attributed to increases in more stable soil aggregates. The protein, glomalin, symbiotically along the roots of legumes and other plants, serves

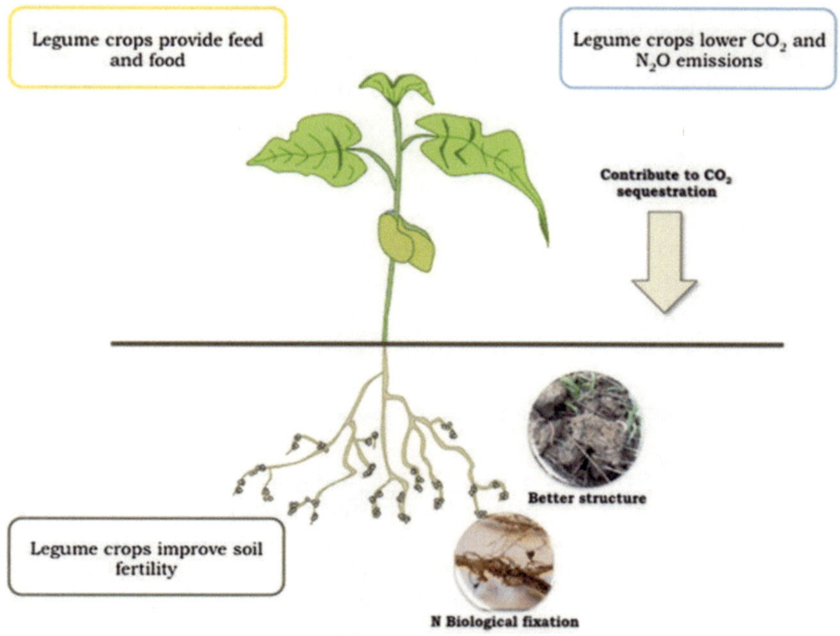

Figure 1.
Benefits of legume crop in improving soil sustainability [10].

as a "glue" that binds soil together into stable aggregates. This aggregate stability increases pore space and tilth, reducing both soil erodibility and crusting.

3.5 Lower soil pH

Because inoculated, nodulated legumes acquire their N from the air as diatomic N rather than from the soil as nitrate, their net effect is to lower the pH of the soil. In greenhouse studies, alfalfa and soybeans lowered the pH in a clay loam soil by one whole pH unit. Legumes could lower the pH and promote increased plant-soil-microbial activity on soils with a pH above the range for optimum crop growth and development.

4. Biological diversity

Legumes contribute to an increased diversity of soil flora and fauna lending a greater stability to the total life of the soil. Legumes also foster production of a greater total biomass in the soil by providing additional N. Soil microbes use the increased N to break down carbon-rich residues of crops like wheat or corn.

4.1 Legumes and carbon sequestration

For a range of years, the practicable importance of legumes in many agro-ecosystems, but also the restrained extent to which this possible has been realized, has been recognized. Legumes do not just contribute in terms of food, feed and fertility, but are also essential as fuelwood and with admire to carbon (C) sequestration. In this chapter we focal point on the extent to which legumes can contribute to greater C sequestration and the delivery of co-benefits including greater biodiversity and reduced greenhouse fuel (GHG) emissions. We additionally consider briefly the main reasons why legumes are currently underutilized and the possibilities for a larger function in the future. Enhancing C sequestration in the soil is linked to elevated biomass and hence to soil fertility. Raising fertility is perchance the most effective way of rapidly growing carbon sink capacity. Clearly, one way of doing this is through elevated addition of nitrogenous fertilizers. However, caution in the enormous use of nitrogenous fertilizers as a strategy to elevated productivity is excellent for a variety of reasons, consisting of the potential for other emissions. By contrast, the role of legumes in supplying nitrogen (N) through fixation is being increasingly more seen as important as an extra beneficial in terms of common GHG stability than had as soon as been thought. The introduction of legumes and their higher utilization as section of a pasture improvement system are consequently probably to be worthy of serious consideration in many circumstances [11].

4.2 Reduction in greenhouse gasoline emissions

Legumes are also possibly to have a position to play in lowering GHG emissions from ruminant systems. An approach to decreasing methane emissions of current interest and supported by some initial evidence is the use of tannin containing forages and breeding of forage species with greater tannin content. Forage legumes such as *Lotus corniculatus* (birds foot trefoil) and *L. uliginosus* (greater trefoil) possess secondary metabolites acknowledged as condensed tannins in their leaves. They are no longer present on the leaves of white or purple clover but are existing in the inflorescences. Methane production values had been lower in housed sheep fed on purple clover and birds foot trefoil than on a ryegrass/white clover pasture [12]. The emissions of nitrous oxide from soils improved linearly with the quantity of mineral nitrogen fertilizer applied and because structures containing legumes produce lower annual nitrous oxide emissions, alfalfa and different legume vegetation need to be regarded differently when deriving national inventories of GHG from agriculture. The nitrous oxide emissions are from soils with alfalfa and soybean cropping, looking at soil floor emissions in evaluation with perennial grass. Low nitrous oxide emissions have been considered under grass and soil mineral N used to be up to ten instances higher beneath legumes however soil mineral N pools were not carefully associated to nitrous oxide emissions. Comparable emissions were viewed under timothy (*Phleum pratense*) as underneath legumes.

Legumes are soil-amendment crops with strong benefits on soil health and need to be an essential element of the farming systems [13]. Legumes have positive effects on soil processes such as benefiting agroecosystems, agricultural productivity, soil conservation, soil biology, SOC and N stocks, soil chemical and bodily properties, BNF, nitrous oxide (N_2O) emission, and nitrate (NO_3) leaching by means of lowering the need for chemical fertilizers. Above all, legumes are now utilized as soil nourishment agents. However, these benefits on soil health need to be quantified, and their mechanisms understood. Thus, incorporating legumes as a section of cropping systems is pertinent to higher soil fitness and productivity [14].

4.3 The potential for legumes to mitigate climate change

The concentrations rise, it has become an increasing number of necessary to account for losses of CO_2 and N_2O arising from agriculture. Emissions of these gases may occur either directly as the result of farming activities (e.g., cultivation and harvesting) or circuitously for the duration of the production and transport of required inputs (e.g., fertilizers, herbicides, and pesticides). The plausible function of N_2-fixing legumes in lowering GHG emissions via direct effects on CO_2 and N_2O fluxes in the production of high-protein grain and forage will be in contrast to the functions of fertilizer N in the following sections. CO_2 emissions bobbing up from N fertilizer manufacturing and symbiotic N_2 fixation.

4.4 Role of legume vegetation on enhancing soil physical properties

Important soil physical properties are bulk density, porosity, combination stability, and texture. These properties are additionally associated with water-related methods including aeration, runoff, erosion, water maintaining capacity, and infiltration rate [15]. Legume vegetation have a manageable to enhance physical properties of soil by being a soil conditioner and enhancing the physical residences [16]. Leguminous cover crops have a tremendous effect on soil physical properties broadly speaking due to the manufacturing ability of large biomass which affords substrata for soil organic undertaking and soil organic matter [17]. Furthermore, leguminous cover vegetation are grown to protect the soil from loss of plant nutrients and erosion, while green manure plants are grown for the motive of improving soil bodily properties. Moreover, some plants can physically modify the types of soil profile. Legumes additionally have an effect on soil shape by means of their impact on aggregation. Leguminous cover crops can expand or keep an appropriate soil C/N ratio and increase in preserving soil organic carbon stock. Legume plants often result in higher infiltration of water, due to direct effects of the crop residue in soil formation and aggregation [18–26].

4.5 Role of legume crops on improving soil chemical properties

Soil chemical properties for sustainability are connected with the capability to provide vitamins for crop and retaining/denaturing hazardous chemical compounds or factors to the agroecosystem. Soil cation alternate capability (CEC), pH, nutrient levels, and soil organic carbon concentration are the primary chemical elements used toward the evaluation of soil fertility. Soil chemical properties have been associated with leguminous crops, and thus, the particulars of a soil property are easily interpreted and permit a rapid enhancement of the soil chemical properties through N-fixation and root biomass. Legume-based rotation induces modifications in the pH of the rhizosphere sector of soil. Root exudation of legumes and change or release of organic acids on the epidermal cell of root surfaces can also enhance P availability [27]. In addition, changes in pH are broadly recognized to affect the increase and undertaking of microorganisms [28], which are additionally necessary aspects in nutrient cycling processes. Leguminous green manure is a well-known generator of soil natural matter. Green manure, apart from increasing soil N, releases P, continues and renews the soil natural carbon, and improves soil chemical characteristics. Incorporation of legume residues is really useful to the soil for growing soil natural carbon awareness which is not only vital to agricultural productiveness however also to sequestration of C from atmospheric CO_2 [29]. Observed that when leguminous cover plants are used as green manure and incorporated into the soil, their residues make bigger availability of N, P, K, and trace elements to the succeeding plants due to the lowering of the soil pH brought about by the CO_2 produced in the process of decomposition [30].

4.6 Role of legume vegetation on enhancing soil microbial biomass

Soil microorganisms have a necessary link between plant productiveness and soil nutrient availability as they are indirectly directly engaged in the nutrients cycling through the conversion of inorganic and organic types of nutrients [31]. Legumes are one of the necessary components to increase soil microbial biomass in soils. Legumes play a necessary function in SMB and energetic key strategies such as nutrient cycling and soil organic matter decomposition and, thus, improve crop productiveness and soil sustainability [32]. Some microorganisms which interact physically with leguminous vegetation in the rhizospheric zone can also enhance crop productivity positively by enhancing plant increase and development [33].

4.7 Role of legume crops in soil carbon sequestration

Sequestration of soil organic carbon is one of the vital determinants of soil fertility, productivity, and quality. Crop residues increase carbon sequestration through decomposition of their residues. Increase in soil natural carbon stock improves soil tilth and workability, stabilizes soil aggregates, will increase soil water preserving and aeration, enhances buffering capacities, and improves availability of nutrients through breakdown of residues [34]. The soil organic carbon inventory depends on soil types, crop and residue management [35], fertilizer N input, and frequency and kind of cropping device [36]. In the agricultural fields, legume plants make contributions positively to the soil natural carbon stock, soil tilth, soil fertility, and universal soil sustainability. Legume-based cropping systems improve mixture balance and lengthen the nutrient dwelling time in soil through decreasing the mineralization rate. Biomass production can be expanded by legume based bi culture, a combination of legume with nonlegume species [8, 37].

4.8 Role of legume vegetation in improving the soil N pool

Nitrogen is vital for the crop growth, solely to water and light. However, most vegetation depend on the consumption of soil N to meet their needs; most highly the legumes, are capable of N-fixation with the symbiotic relationship with rhizobia. The BNF benefits not only the legumes however also improves yield in succeeding crops, in agroforestry systems, and in legume-cereal intercropping system. The N quantity made available to cereal crop derives from the breakdown of legume-biomass residues. However, the affiliation of N tends to cross from crop containing enormously high N (i.e., legumes) to those with an increased N demand (nonlegume). An approach to raise N supply in cropping structures is the inclusion of N-fixing leguminous crops, which can grant N advantages to the vegetation thru N transfer. The extent of biologically fixed N/year by way of legumes varies significantly from zero to several hundred kg N/ha.

4.9 Role of legumes in mitigating greenhouse gas and enhancing soil pleasant

1. Lower the emission of greenhouse gases (GHG) such as carbon dioxide (CO_2) and nitrous oxide (N_2O) compared with agricultural systems based totally on mineral N fertilization.

2. Have an essential role in the sequestration of carbon in soils.

3. Reduce the overall fossil power inputs in the system.

5. Importance of legumes

Increased cultivation of legumes is integral for the regeneration of nutrient-deficient soils and for imparting wanted protein, minerals, and nutritional vitamins to human beings and livestock. Legumes can be an ability of improving the livelihoods of smallholder farmers round the world.

5.1 Legumes in human nutrition

- As a supply of protein, grain legumes (such as pigeon pea, chickpea, soybean or mung bean) are a true supply of protein, with a protein content material ranging from 17 to 40%.

  By combining cereal and grain consumption, farmers and their families can achieve protein stability and dietary improvement.

- As a supply of essential vitamins and minerals, legume seeds contain tremendous quantities of minerals (calcium, zinc, iron) and nutritional vitamins (folic acid and diet B).

5.2 Legumes for animal nutrition

Cereal crop residues supplemented with forage legumes notably increase normal animal productivity. For example, improved fowl egg production has been mentioned when pulse grains are protected in their feed. Adding the residue from legume flora into cattle forage can expand the digestibility and typical quality of cereal crop residues. For example, maize residues tend to be high in carbohydrates however low in protein; therefore, adding leguminous flora will make a contribution to multiplied livestock nutrition.

5.3 Legumes for crop and soil improvement

For most effective yield, plants require a furnish of mineral nutrients, the most essential of which is nitrogen. Exhausted soils are often low in nitrogen, meaning that farmers are usually applying inorganic fertilizers. However, as fertilizer expenses increase, farmers battle to acquire properly yields. This trouble can be addressed by incorporating legumes into the cropping system. Leguminous plant have a close relationship with nitrogen-fixing microorganism known as Rhizobium. By biologically fixing nitrogen ranges in the soil, legumes grant a fantastically low-cost approach of changing nitrogen in the soil, improving soil fertility and boosting subsequent crop yields.

6. Conclusion

The use of nitrogen-fixing legume-based leys, whether they are used for grazing, conservation or mulched to build soil fertility, is the basis of most organic systems. Their use is enshrined in the organic standards, which require the inclusion of legumes in rotations. The wider benefits of legumes, particularly in providing food for pollinators, are also increasingly being recognized. Globally, the amount of carbon di oxide respired from the root systems of N_2-fixing legumes could be comparable to, or higher than, the carbon di oxide generated during nitrogen-fertilizer production. However, the carbon di oxide respired from the nodulated roots of legumes originated from the atmosphere via photosynthesis, so any of the carbon di oxide that was not subsequently recaptured by the plant and eventually escaped from the legume canopy to the atmosphere would essentially be carbon neutral. By contrast, all the carbon di oxide released during the synthesis of fertilizer nitrogen would be derived from fossil energy and represents a net contribution to atmospheric concentrations of carbon di oxide.

Author details

Muthuraman Yuvaraj[1*], Muthaiyan Pandiyan[1] and Pandurangan Gayathri[2]

1 Agricultural College and Research Institute, Tamil Nadu, India

2 Adhiparasakthi Agricultural College, Tamil Nadu, India

*Address all correspondence to: yuvasoil@gmail.com

References

[1] United Nations: World population prospects: The 2012 revision, key findings and advance tables. In: Working Paper No. ESA/P/WP.227. New York: United Nations, Department of Economic and Social Affairs, Population Division; 2013

[2] Yadav SS, Hunter D, Redden B, Nang M, Yadava DK, Habibi AB. Impact of climate change on agriculture production, food, and nutritional security. In: Redden R, Yadav SS, Maxted N, Dulloo MS, Guarino L, Smith P, editors. Crop Wild Relatives and Climate Change. New Jersey, USA: Wiley; 2015. pp. 1-23

[3] Voisin AS, Gueguen J, Huyghe C, Jeuffroy MH, Magrini MB, Meynard JM, et al. Legumes for feed, food, biomaterials and bioenergy in Europe: A review. Agronomy for Sustainable Development. 2014;**34**:361

[4] Tharanathan RN, Mahadevamma S. Grain legumes a boon to human nutrition. Trends in Food Science and Technology. 2003;**14**:507-518

[5] Lemke RL, Zhong Z, Campbell CA, Zentner RP. Can pulse crops play a role in mitigating greenhouse gases from North American agriculture? Agronomy Journal. 2007;**99**:1719-1725

[6] Mtambanengwe F, Mapfumo P. Combating food insecurity on sandy soils in Zimbabwe: The legume challenge. Symbiosis. 2009;**48**:25-36

[7] Mytton LR, Cresswell A, Colbourn P. Improvement in soil structure associated with white clover. Grass and Forage Science. 1993;**48**:84-90

[8] Naab JB, Chimphango SMB, Dakora FD. N_2 fixation in cowpea plants grown in farmers' fields in the Upper West Region of Ghana, measured using 15N natural abundance. Symbiosis. 2009;**48**:37-46

[9] Deakin WJ, Broughton WJ. Symbiotic use of pathogenic strategies: Rhizobial protein secretion systems. Applied Soil Ecology. 2009;**7**:312-320

[10] Stagnari F, Maggio A, Galieni A, et al. Multiple benefits of legumes for agriculture sustainability: An overview. Chemical and Biological Technologies in Agriculture. 2017;**4**(2). DOI: 10.1186/s40538-016-0085-1

[11] Dexter AR. Soil physical quality. Part I. Theory, effects of soil texture, density, and organic matter, and effects on root growth. Geoderma. 2004;**120**:201-214

[12] Dhakal Y, Meena RS, De N, Verma SK, Singh A. Growth, yield and nutrient content of mung bean (*Vigna radiata* L.) in response to INM in eastern Uttar Pradesh, India. Bangladesh Journal of Botany. 2015;**44**(3):479-482

[13] Dhakal Y, Meena RS, Kumar S. Effect of INM on nodulation, yield, quality and available nutrient status in soil after harvest of green gram. Legume Research. 2016;**39**(4):590-594

[14] Hauggaard-Nielsen H, Jornsgaard B, Kinane J, Jensen ES. Grain legume–cereal intercropping: The practical application of diversity, competition and facilitation in arable and organic cropping systems. Renewable Agriculture and Food Systems. 2007;**23**:3-12

[15] Juma KA, Averbeke V. Response of Muxe to N and P availability in pots. In: Tenywa JS, Adipala E, Nampala P, Tussiime G, Okori P, Kyanmuhangire W, editors. African Crop Science. Proceedings; , vol. 7, Kampala, Uganda, 5-9 December 2005. pp. 1179-1182

[16] Lal R. Climate change and soil degradation mitigation by sustainable

management of soils and other natural resources. Agricultural Research. 2012;**1**:199-212

[17] Lal R. Restoring soil quality to mitigate soil degradation. Sustainability. 2015;7:5875-5895

[18] Liang B, Lehmann J, Sohi SP, Thies JE, O'Neill B, Trujillo L, et al. Black carbon affects the cycling of non-black carbon in soil. Organic Geochemistry. 2010;**41**:206-213

[19] McLauchlin KK, Hobbie SE. Comparison of labile soil organic matter fractionation techniques. Soil Science Society of America Journal. 2004;**68**:1616-1625

[20] Meena RS, Meena PD, Yadav GS, Yadav SS. Phosphate solubilizing microorganisms: Principles and application of Microphos Technology. Journal of Cleaner Production. 2017;**145**:157-158

[21] Meena RS, Meena VS, Meena SK, Verma JP. The needs of healthy soils for a healthy world. Journal of Cleaner Production. 2015c;**102**:560-561

[22] Mousavi S, Yousefi-Moghadam S, Mostafazadeh-Fard B, Hemmat A, Yazdani MR. Effect of puddling intensity on physical properties of a silty clay soil under laboratory and field conditions. Paddy and Water Environment. 2009;7(1):45-54

[23] Pikul JL, Johnson JMF Jr, Schumacher TE, Vigil M, Riedell WE. Change in surface soil carbon under rotated corn in eastern South Dakota. Soil Science Society of America Journal. 2008;**72**:1738-1744

[24] Schimel DS, Braswell BH, Holland EA. Climatic, edaphic and biotic controls over storage and turnover of carbon in soils. Global Biogeochemical Cycles. 1994;**8**:279-293

[25] Srinivasarao C, Venkateswarlu B, Lal R. Long-term effects of soil fertility management on carbon sequestration in a rice-lentil cropping system of the Indo-Gangetic plains. Soil Science Society of America Journal. 2012;**76**(1):167-178

[26] Turnbull J, Bowman WD. Variable effects of nitrogen additions on the stability and turnover of soil carbon. Nature. 2002;**419**:915-917

[27] Varma D, Meena RS, Kumar S, Kumar E. Response of mung bean to NPK and lime under the conditions of Vindhyan Region of Uttar Pradesh. Legume Research. 2017;**40**(3):542-545

[28] Webb J, Bellamy P, Loveland PJ, Goodlass G. Crop residue returns and equilibrium soil organic carbon in England and Wales. Soil Science Society of America Journal. 2003;**67**:928-936

[29] Ayarza M, Barrios E, Rao IM, Amezquita E, Rondón M. Advances in improving agricultural profitability and overcoming land degradation in savanna and hillside agroecosystems of tropical America. In: Bationo A, Waswa B, Kihara J, Kimetu J, editors. Advances in Integrated Soil Fertility Research in Sub-Saharan Africa: Challenges and Opportunities. The Netherlands: Springer; 2007. pp. 209-229

[30] Benchaar C, Pomar C, Chiquette J. Evaluation of dietary strategies to reduce methane production in ruminants: A modelling approach. Canadian Journal of Animal Science. 2001;**81**:563-574

[31] Douxchamps S, Rao IM, Peters M, van der Hoek R, Schmidt A, Martens S, et al. Trade-off analysis of tropical legumes in small-holder crop-livestock systems in the hillsides of Nicaragua: The case of *Canavalia brasiliensis*. Agricultural Systems. 2013 (in review)

[32] Graham PH, Vance CP. Nitrogen fixation in perspective: An overview of research and extension needs. Field Crops Research. 2000;**65**:93-106

[33] Israel DW. Investigation of the role of phosphorus in symbiotic dinitrogen fixation. Plant Physiology. 1987;**84**:835-840

[34] Jeuffroy MH, Ney B. Crop physiology and productivity. Field Crops Research. 1997;**53**:3-16

[35] Franke AC, Laberge G, Oyewole BD, Schulz S. A comparison between legume technologies and fallow and their effects on maize and soil traits, in two distinct environments of the West African savannah. Nutrient Cycling in Agroecosystems. 2008;**82**:117-135

[36] Rufino MC, Rowe EC, Delve RJ, Giller KE. Nitrogen cycling efficiencies through resource-poor African crop–livestock systems. Agriculture, Ecosystems & Environment. 2006;**112**:261-282

[37] Ladha JK, Pathak H, Krupnik TJ, Six J, Van Kessel C. Efficiency of fertilizer nitrogen in cereal production: Retrospects and prospects. Advances in Agronomy. 2005;**87**:85-156

Chapter 3

Soybean Production, Versatility, and Improvement

Zachary Shea, William M. Singer and Bo Zhang

Abstract

Soybean is one of the most widely planted and used legumes in the world due to its valuable seed composition. The many significant agronomic practices that are utilized in soybean production are highlighted with an emphasis on those used during the pregrowing season and growing season. The various pests of soybeans and the pest management strategies used to control them are described with special attention to insects, weeds, bacteria, fungi, and nematodes. The multitude of soybean uses for livestock and human consumption, and its industrial uses are discussed in this chapter. Additionally, the conventional breeding and genetic engineering attempts to improve soybean protein, oil, and sucrose content as well as eliminate the antinutritional factors, such as trypsin inhibitors, raffinose, stachyose, and phytate, are examined. In this chapter, the various management practices, uses, and breeding efforts of soybean will be discussed.

Keywords: agronomic practices, pest management, soybean uses, breeding, genetic engineering

1. Introduction

Soybean (*Glycine max*) is one of the most valuable, versatile, and nutritionally important legumes globally. It can be grown in a multitude of environments, using a variety of management practices, and for diverse end-user purposes. In 2018, roughly 398 million tons of soybeans were produced worldwide which accounted for 61% of overall oilseed production and 6% of the world's arable land use [1–3]. The United States, Brazil, and Argentina constituted approximately 81% of international soybean production, producing 34, 32, and 15%, respectively [4, 5]. Soybean seed composition and its main components, meal and oil, are the driving forces behind crop production that has increased nearly 350% since 1987 [5]. Soybean meal is intricately connected to the food supply through direct food consumption and indirect consumption as a large source of livestock feed. Soy oil provides great versatility with uses in food and beverage, wax, construction, cosmetics, plastics, and fuel.

Soybean originated in East Asia and has been cultivated in China for millennia. It is estimated that the domestication event from wild soybean (*Glycine soja*) occurred during the Shang Dynasty, 1700–1100 B.C. [6]. While no longer the largest producer, China and other Asian countries continue to incorporate large quantities of traditional and innovative soy foods into their diet. In 2018, China was the largest customer for United States whole soybeans, importing over $3 billion worth [7]. The United States and western countries mainly utilize soybean indirectly in the food supply as livestock feed and food ingredients such as textured vegetable protein and protein isolates. However, as more consumers are looking for plant-based protein in their diet, soy foods will become a globally viable alternative to animal protein. As the soybean appetite has increased and transformed, scientific developments have also improved soybean production through agronomic, management, and genetic methods to meet demand.

2. Agronomic practices

2.1 Pregrowing season practices

Soybeans are one of the most flexible crops in terms of production methods, geographical growing regions, and end use versatility. Therefore, there are multiple agronomic practices to consider when preparing a field for soybean production. While tillage and fertilization practices are common among producers, technique specifications can vary greatly due to preferences, environmental conditions, and cost. Historically, mechanized and non-mechanized tillage was considered a vital practice to maximize crop yield and value [8]. While tillage is still a useful tool, contemporary research has corroborated the dangers of over-tilling and the potential benefits from soil conservation and no-till operations. No-till practices and conservation tillage for soybean are wide-spread in areas of highly-erodible soil, and some research has shown that soybean yields remain the same or increase with decreased tillage [9–12]. However, other research has shown that rotational tillage practices will provide higher crop value than no-till practices, specifically because of herbicide costs and equipment requirements [13–15]. Given the need for proper soil maintenance, conservation tillage (<30% crop residue left on the soil surface) is a popular compromise, especially in herbicide tolerant soybean production [16, 17]. Research has further elucidated the benefits of conservation tillage and no-till practices on soil health by showing positive correlations with rhizobia and nematode populations [18, 19].

Pre-plant fertilization for a variety of macro and micronutrients is another common practice in soybean production. Soil fertility programs are designed to provide sufficient nutrients for a crop's needs which maximizes crop yield and farm efficiency while also minimizing environmental impact. To prepare a field for soybean planting, a farmer must start by determining what nutrients are already present in the soil; this can be accomplished by a variety of soil sampling and analysis methods [20]. The primary macronutrients, nitrogen (N), phosphorous (P), and potassium (K), should be examined first alongside critical secondary macronutrients and micronutrients such as sulfur (S), calcium (Ca), magnesium (Mg), zinc (Zn), manganese (Mn), boron (B), iron (Fe), and copper (Cu). General field nutrient requirement guidelines for soybean production are summarized in **Table 1**. While soybeans require a large amount of nitrogen, fertilization is usually unnecessary because of the symbiotic relationship with *Bradyrhizobium japonicum*,

	N	P_2O_5	K_2O	Ca	Mg	S	Zn	Mn	B	Fe	Cu
kg/ha	275	48	207	113	50	19	0.34	0.37	0.33	0.85	0.06
lb/ac	245	43	185	101	45	17	0.3	0.33	0.29	0.76	0.05

Table 1.
Estimated nutrient uptake and accumulation for 3500 kg/ha^{-1} (52 bu/ac) soybean yield [21–23].

a bacterium that performs nitrogen fixation and provides plant available nitrogen [9, 21, 22]. Depending on field conditions, 25–75% of nitrogen in mature soybeans can originate from symbiotic nitrogen fixation [24]. Excess nitrogen has been linked to negative plant physiological conditions and inhibited rhizobia activity [25]. Thus, the best solution to limited field nitrogen is sometimes soil or seed-applied bacterial inoculation [21, 22, 25]. Soil pH is also a vital component of field management. It is well-documented that all nutrients have varying availability to plants depending on pH [22, 26, 27]. Generally, soybeans prefer a slightly acidic soil ranging from 6 to 7 pH [9, 22]. Liming a field is the optimum technique to raise pH, while the most common practice for lowering pH is elemental sulfur application.

While yield is the driving factor for fertilization, recent market changes have adjusted soybean valuation with increased focus on seed composition quality. Amino acid profiles as descriptors for protein quality in human food and livestock feed as well as high oleic acid soybeans for increased functionality and performance are just two examples of possible premiums producers can receive through soybean seed composition. Research has shown that agronomic practices coupled with location-dependent, environmental variables can directly impact those premiums [28–30]. Nitrogen fertilization plays a limited role in seed composition as it is rarely needed due to the bacterial nitrogen fixation. However, excess nitrogen has been shown to decrease the levels of sulfur-containing amino acids and has an inconsistent effect on fatty acid concentrations [31, 32]. Phosphorous applications can increase protein quantity without adjusting the amino acid profile but also has a positive correlation with higher phytic acid and isoflavone concentrations [33–35]. Additionally, phytic acid has been shown to increase alongside zinc concentrations [33]. Pre-plant potassium applications have limited return on investment in regard to yield and seed composition; however, potassium deficient soybean plants are at a greater risk to insect pests, specifically aphids [36, 37]. Limited yield response is observed with sulfur applications. Although researchers have found the use of sulfur fertilizers to be economically viable, particularly on coarse soils, sulfur fertilization is rarely recommended [9, 21, 38]. Soil sulfur levels have also been shown to greatly impact the ratio between 11S and 7S seed storage proteins [39]. As markets continue to change and value differing soybean seed compositions, it will be critical for producers to fertilize with both yield and seed components in mind.

2.2 Growing season practices

Soybean producers make decisions throughout the year that impact final yield, value, and profit from their annual crop. Many of the most critical decisions occur at the beginning of and throughout the growing season. From the moment a soybean seed is planted to harvest, producers choose (or decide against) a multitude of practices including crop rotation, row spacing, population density, irrigation, post-emergence fertilization, and pest management. Maximizing a potential soybean crop is directly connected to previous field usage. Crop rotation or the process of growing different crops in sequenced seasons within the same field is a common practice in soybean production. Corn (*Zea mays*) and soybean rotations are advantageous because of corn's high nitrogen demand which can be alleviated through *Bradyrhizobium japonicum* nitrogen fixation in soybean nodules. Corn and soybean rotations also exhibit beneficial energy balance and grain yield improvement [13, 40]. Rotations including corn and soybeans as well as wheat (*Triticum*), oats (*Avena sativa*), barley (*Hordeum vulgare*), cotton (*Gossypium*), and forageable pasture have also shown potential for economic and environmental gains [22, 41, 42]. Although depending on crop sequences, new management practices may be needed. For example, alfalfa or clover following soybeans would require liming for maximum production as those crops prefer a slightly higher pH [43].

Protecting and revitalizing the soil through non-harvested crops planted between soybean growing seasons or cover cropping is also beneficial. Cover crops protect the soil that would otherwise be farrow and replenish nutrients assimilated into the soybean plant [9, 22, 43]. Furthermore, cover crops can beneficially reduce weed pressure, lessen soil compaction, and improve water conservation [44–46]. However, cover crops increase annual cost and have not been shown to increase soybean yield which can negatively impact certain producer's net profit [47]. Many farmers who receive enough growing degree units throughout the year also limit farrow fields by double cropping with soybean. Soybean and wheat double crop systems have exhibited high economic returns for producers in both field and modeling research [48, 49]. Double cropped soybeans exhibit lower yield due to late planting and decreased leaf-area-index potential, but this can be mitigated with early maturing varieties [50, 51]. Intercropping or growing at least two crops simultaneously is another, less-common option for soybean production. Corn and soybean intercropping can increase yields for both crops with the proper seeding rates [52, 53]. Wheat and soybean intercropping also displays positive yield response [54, 55]. Sugarcane (*Saccharum officinarum*) and soybean interspecific relationships increase sugarcane yield and improves rhizospheric activity while reducing soybean yield [56]. While intercropping can enhance value for soybean producers, it is unsuitable for most large-scale production systems.

After choosing a cropping system, soybean producers must then determine the proper row spacing and population density for their environment. The appropriate balance between row space and plant density is critical for maximum soybean production and reliable economic returns. Narrow rows and high plant densities both correlate with quickened canopy closure and weed suppression [57–60]. Increased plants per field also increase cost; however, subsequent increased yield and profit overcomes the cost [61–63]. As soybeans emerge and grow, the next consideration for producers is irrigation. This localized decision can be based upon historical precipitation records, predicted forecasts, day-to-day weather events, or a combination of factors. In the absence of natural precipitation, irrigation is vital to soybean production as water deficiencies inhibit yield potential [64–67]. Irrigation can also be optimized spatially throughout a field with variable rate techniques and temporally across the growing season by targeting specific growth stages [68, 69]. Fertigation applications can be used to combine applications of post-emergent fertilizer with irrigation. Other methods of post-emergent fertilization including foliar spray and direct-to-soil applications are more common solutions for growing season nutrient issues. Plant tissue sampling and analysis can be coupled with soil samples to determine in-season soil deficiencies and to prescribe further applications [9, 21, 26]. As soybean increases nitrogen uptake during reproductive stages when bacterial fixation may be diminishing, soil or foliar nitrogen applications are typical yet usually ineffective. While limited yield increases can be seen from supplemental nitrogen applications or various nutrient combinations, the economic returns generally fail to cover the cost of application [70–73]. Foliar nutrient applications have shown minor impacts on seed protein and oil content; however, these results are inconsistent amongst experiments [74, 75]. A location-specific, comprehensive nutrient management plan that accounts for all other agronomic practices is the best method for maximizing yield and economic returns in soybean production.

3. Pest management

3.1 Insect pests

Insect and insect-like pests of soybean vary greatly ranging from aphids to stinkbugs to loopers to beetles. Which insects are the major pests and potential pest impact on soybean varies significantly from year to year and depends on the region the soybean crop is grown. Total damage by insects is a little ambiguous but yield losses of up to 80% have been reported [76]. Some prominent insect pests include soybean aphids (*Aphis glycines*), Japanese beetle (*Popillia japonica* Newman), Mexican bean beetle (*Epilachna varivestis* Mulsant), two-spotted spider mites (*Tetranychus urticae*), brown marmorated and red banded stinkbug (*Halyomorpha halys* and *Piezodorous guildinii*), bean leaf beetle (*Cerotoma trifurcata*), and kudzu bug (*Megacopta cribraria* Fabricius) [76–81].

Insecticides constitute a large portion of insect management as they are used to control most insect pests and in some cases are the primary method of control [80]. Integrated pest management (IPM) is becoming more common among growers due to its ability to reduce pesticide use, non-pests affected, workers' exposure to pesticides, and the likelihood insecticide resistance [82, 83]. Additionally, it has been found to be effective at reducing damage done by pests equivalent to conventional methods [84]. IPM works similarly for all pests. It involves monitoring fields to determine which pests are present, determining which pesticides can and should be used, and incorporating cultural management practices [83]. For insects, trap cropping and sweep nets are used to monitor and determine which insect pests are present [85, 86]. The cultural practices used in insect management include altering planting date and row spacing, using no-till fields, and using resistant soybean cultivars [76–81].

3.2 Weeds

Weeds are considered one of the most damaging, if not the most damaging pests, in soybean [87]. About 37% of global production of soybean is affected by soybean, while 23% of global production is affected by other pests [88]. In the United States alone, it has caused losses of several million US dollars each year [87]. Weeds pose a problem for soybean crops since they compete for nutrients, space, and other resources [89]. There are many different weed pests that compete with soybean, some of which include common waterhemp (*Amaranthus rudis*), Canadian horseweed (*Conyza canadensis*), giant ragweed (*Ambrosia trifida*), ivy-leaf morning glory (*Ipomea hederacea*), common cocklebur (*Xanthium strumarium*), Johnsongrass (*Sorghum halepense*), and pigweed (*Amaranthus* spp.) [90, 91]. It is important to note that which weeds are found in a particular field depends largely on where the soybean crops are grown.

Management of weeds is largely done through integrated pest management. This involves using herbicides along with herbicide resistant soybean varieties and cultural practices [87, 92]. There are many different classes of herbicides that include enzyme inhibitors, lipid synthesis inhibitors, photosystems diverters, nucleic acid inhibitors, and auxin inhibitors [93]. Historically, herbicides have been a large part of weed management and will most likely remain significant due to effectiveness and limited efficiency through other individual methods [87]. Furthermore, herbicide effectiveness can be improved by using herbicide resistant soybean, such as glyphosate resistant Roundup Ready soybean. Although since weeds can develop resistance to herbicides, it is important to incorporate other management practices [87]. One such method is herbicide spray timing. A common management practice involves pre- and postemergence herbicide applications. This involves spraying herbicides before and a few days after the soybean plants have emerged to reduce any damage to the soybean plants [94]. Additionally, cultural control practices are used including crop rotations, planting in narrow rows and proper fertilization to promote crop competition, and cultivation [92]. Crop rotations allow for different herbicides to be used which in turn helps to prevent the development of herbicide resistant weeds [92]. Promoting crop competition through planting density allows soybean plants to grow enough to create a canopy to maximize shading of weeds [92]. Cultivation is an effective and economical way to control weeds to help minimize herbicide use [92]. All of the aforementioned management practices are parts of integrated weed management and will continue to play a significant role in control of weeds.

3.3 Diseases

Similar to the insect pests, there is a wide variety of diseases in soybean. Most diseases are caused by fungal and bacterial diseases and can be vectored by nematodes. Fungal diseases have been known to reduce yield up to 50%, while bacterial diseases have been known to cause yield loss of anywhere between 15 and 60% [76]. Which disease is the most devastating depends on the region and the year, but the most prevalent diseases include *Heterodera glycines*, *Phytophthora sojae*, *Colletotrichum truncatum*, *Septoria glycines*, and *Phakopsora pachyrhizi* [76]. Of these five diseases, *Heterodera glycines*, or soybean cyst nematode is the most economic damaging disease being found in all countries that grow soybean and causing up to 90% yield reduction in some areas [76]. **Table 2** provides an overview of some of the main soybean diseases.

From **Table 2**, it is evident that chemical pesticides still play a large role in treatment strategies against all major diseases in soybean. However, there has been a rising interest to incorporate other methods that prevent and treat diseases in soybean due to the harmful environmental and health effects of pesticides. Some other methods to control soybean diseases are seen in cultural control practices, such as increasing or decreasing tillage and crop rotation, drainage, and using resistant cultivars [103]. While the treatments listed in the above table are usually effective, there is continual research to find innovative ways to improve the control of plant diseases. One such example is the development of using hyperspectral bands for early detection of charcoal rot in soybean [104]. These researchers developed a method that involves analyzing spectral and spatial information of infected and healthy soybean in order to find wavebands that signify a soybean plant that is infected with charcoal rot [104]. This process identified six wavebands that were specific to plants infected with charcoal rot and can potentially allow for the detection of charcoal rot in crops in three days [104]. By being able to identify disease earlier, growers can minimize the damage done by that disease by removing infected plants and incorporating treatment strategies, such as pesticides or cultural controls.

The research above shows that there is interest in developing early detection for soybean pathogens. One of the other major areas of research for soybean diseases, is identifying resistance genes to promote resistant cultivars. Given that soybean cyst nematode is one of the most devastating soybean diseases there has been a lot of research done to identify genes involved with resistance to soybean cyst nematode. The main resistance gene in soybean to cyst nematode is the Rhg1 gene, which encodes an amino acid transporter [105, 106]. This gene confers partial resistance and has been shown to reduce reproduction of soybean cyst nematode and improve yield in fields that are infected with soybean cyst nematode [78]. Even though there are resistant cultivars available, they do not permanently stop diseases. For soybean cyst nematode, it is advised to utilize cultural practices, such as using multiple resistant cultivars and rotating with non-host crops that are resistant to cyst nematode, and other methods [107]. This is a classic example of how integrated pest management involves continuously incorporating new methods to control diseases to prevent the disease from overcoming any pesticides and resistant cultivars.

Name	Type	Causal Agent	Transmission	Symptoms	Treatments
Charcoal rot	Fungal	*Macrophomina phaseolina*	Soil born	Wilting, necrosis, black/dusty microsclerotia on stem/pods/seeds, brown lesions on emerged seedlings	Fungicides, resistant cultivars, reduce tillage, crop rotations
Soybean Cyst Nematode	Nematode	*Heterodera glycines*	Soil born	Stunted roots, can increase sensitivity to some fungal diseases, presence of cysts on roots	Resistant cultivars, crop rotations
Phytophthora Root and Stem Rot	Oomycete	*Phytophthora sojae*	Can overwinter in soil, water	Reddish-brown/black lesions on stem, chlorotic leaves, soft rot on roots, seed rot, emergence damping off of seedlings	Seed treated fungicides, improving soil drainage, resistant cultivars
Soybean Bacterial Blight	Bacterial	*Pseudomonas syringae pv. glycinea*	Water, can overwinter in plant residue	Affects mid to upper leaves, yellowish-brown, angular lesions on leaves, discolored/shriveled seeds, water soaking	Copper fungicides, resistant cultivars, increasing tillage, crop rotations
Soybean Anthracnose	Fungal	*Colletotrichum spp*	Seedborn, can overwinter in plant residue	Brown lesions with setae, pods with fewer seed, brown cankers, defoliation, damping off	Crop rotation, fungicide treated seed
Brown spot/ Septoria Brown Spot	Fungal	*Septoria glycines*	Can be transmitted through infected seed, can overwinter in plant residue	Small, brown lesions on leaves, yellowing leaves, lesions contain pycnidia	Foliar fungicides, crop rotation, increased tillage
Soybean Rust	Fungal	*Phakospora pachyrhizi*	Spores are spread by wind	Reddish/brown lesions with pustules on leaves, pods, and seeds	Fungicides

Table 2.
Overview of 7 prevalent diseases in soybean [95–102].

4. Soybean utilization and products

4.1 Livestock feed

Soybean is a valuable crop worldwide mainly because of soybean meal's nutritional efficacy as a food and feed ingredient. A high protein content, balanced essential amino acid profile, and the presence of other beneficial nutrients all contribute to its economic and nutritional value. Soybean meal constitutes 70% of seed value while only being roughly 35% of seed dry weight [108, 109]. Furthermore, in the United States, 97% of soybean meal is used for livestock feed [109]. This overwhelming usage rate as a livestock protein source is mainly due to the presence of essential amino acids. While some livestock require other amino acids, most livestock need nine essential amino acids: histidine, isoleucine, leucine, lysine, methionine, phenylalanine, threonine, tryptophan, and valine [110]. All nine of these amino acids are found in some quantity in soybean meal [111]. For this reason, soybean meal can maximize livestock production in cattle, swine, poultry, and aquaculture. Generally, soybean meal and other soy byproducts use are limited to a supplementary or finishing role for cattle due to feed ration complications from other seed components [110, 112, 113]. Soybean meal use is highly prevalent in monogastric livestock production such as swine and poultry and is increasing in popularity for aquaculture [114, 115]. However, soybean as feed has two main obstacles: methionine deficiency and trypsin inhibitor proteins. Albeit present in soybean, methionine content is deficient for livestock needs, is considered the first limiting amino acid for soybean meal and requires producers to supplement with synthetic methionine [116–120]. This has a variety of negative economic and environmental impacts, including increased cost and poor nitrogen-use efficiency [121, 122]. Trypsin inhibitor proteins are an antinutritional factor present in raw soybean that decreases feed efficiency and can harm the livestock. There are a variety of industrial processing methods used to overcome trypsin inhibitors in soybean such as thermal and infrared treatment [123]. In the future, soybean methionine deficiencies and trypsin inhibitor levels may both be solved via breeding and transgenic efforts.

4.2 Human food

Soybean as human food exists to two different extents that are derived from geography and cultural tradition. Eastern hemisphere populations incorporate whole soybeans and processed soy foods into their lives on a daily basis, whereas Western hemisphere populations generally utilize processed soybeans as food ingredients. Eastern soy foods are divided into two main categories: fermented and non-fermented. Non-fermented soy foods include whole seed options such as whole dry soybeans, soy nuts, and edamame, processed items such as soy flour and soy milk, and vegetative soy sprouts [115, 124]. Soy milk in its simplest form is a water extract from soybean that when further processed can make tofu and tofu byproducts such as okara (soy pulp) and yuba (tofu skin). Fermented soy products include miso, soy sauce, tempeh, natto, and sufu, and each product has a specific bacterial species that enables proper fermentation. For example, natto is associated with Bacillus subtilis, and soy sauce is associated with Aspergillus sp. [115, 124]. Western cultures have assimilated many soy food products, and they are becoming more popular as consumers seek plant-based protein sources. However, the vast majority of soybeans in western diets consists of food ingredients made from soybean meal and soy oil. Soybean meal can be processed into ingredients such as soy flour, protein concentrates, and protein isolates that are used in bakery mixes, breakfast cereals, baby food, and exercise supplements [114]. Soybean oil is widely used in vegetable oil and margarine mixes for a variety of cooking purposes. The importance of traditional and innovative soy food uses has perpetuated because of the potential health benefits from soy consumption. Soy foods have been shown to play a role in chronic human disease prevention for conditions such as heart disease, osteoporosis, and cancer [125–127]. However, isoflavones, one of the most common seed components linked to disease prevention, is also negatively linked to hormonal health as a phytoestrogen. While large population subsets are concerned about isoflavones negatively impacting fertility, summarized data has shown inconsistent results [128–130]. As consumers continue to seek plant-based protein, soybeans will be the premier source for historically and culturally significant recipes as well as healthy, novel animal meat alternatives.

4.3 Industrial uses

Even though soybean is classified as an oilseed, soybean oil has historically been an afterthought for soybean producers and processors. When markets for soybean meal would falter, researchers and other stakeholders would turn to soybean oil for added value or seek alternative uses for meal components. Modern sustainability and industrial goals have stimulated soy-based product usage in a variety of fields, as summarized in **Table 3**. Soybean oil as biodiesel has experienced the largest

Soybean Oils		Soybean Lecithin	Soybean Powders
Anti-static agents	Lubricants	Alcohol	Adhesives
Candles	Metal casting	Concrete	Antibiotics
Caulks	Oiled fabrics	Inks	Asphalt
Concrete	Paints	Magnetic tapes	Fermentation aids
Crayons	Pesticides	Paint	Packing films
Dust control agents	Plastics	Paper	Firefighting foams
Electrical insulation	Printing inks	Pesticides	Inks
Epoxy	Putty	Pharmaceuticals	Leather substitutes
Fatty acids	Soaps and detergents	Synthetic rubber	Particle boards
Fatty alcohols	Solvents	Softening leather	Pesticides
Fuel	Vinyl	Yeast	Pharmaceuticals
Hydraulic fluids	Wallboard		Plastics
Pesticides			Polyester
Linoleum backing			Textiles

Table 3.
Industrial uses for soybean products [114, 115].

growth with United States consumers using over 2 billion gallons in 2017 [131]. Current biodiesel production methods can create soy-based fuel that perform nearly equal or equivalent to standard diesel fuels and have the potential to become a truly renewable resource when coupled with sustainable farming practices [132–134]. Constantly improving processing methods will continue to augment soybean seed component versatility and create new opportunities for soy-based products.

5. Soybean seed composition improvement

5.1 Breeding efforts

Soybean seed has many beneficial traits, such as high protein, oil, and soluble sugar content [135]. While soybean seed value is defined by these favorable qualities, past and present breeding attempts have sought further improvement. With regards to protein content, breeders have worked with soybean to increase total protein content as well as the amount of sulfur containing amino acids, methionine and cysteine [135]. Methionine and cysteine are of interest since the seed protein is naturally deficient, and these two amino acids can improve the nutritional value of soybean meal [135, 136]. However, some research indicates that total protein content is negatively correlated with other favorable seed qualities, including yield, oil content, and potentially methionine and cysteine content [137, 138]. So far, most breeding efforts to improve protein quality have involved identifying quantitative trait loci (QTL) that are associated with the amino acid content [139]. QTLs are regions of DNA that are associated with a particular trait and allow breeders to select for particular cultivars that have the trait of interest [140]. The composition of soybean seed oil primarily includes linolenic, steric, palmitic, linoleic, and oleic acid [141]. There have been breeding attempts mainly to increase oleic acid in soybean seed while keeping linolenic acid relatively low, due to respective human health impacts [141, 142]. Lastly, soluble sugar levels, specifically sucrose content, has also been an area of interest in soybean breeding [135]. While sucrose is the main sugar found in soybean, fructose and glucose are also present but in trace amounts [143]. Similar to protein content, multiple QTLs have been identified associated with high sucrose [144]. Sucrose is a desirable seed composition trait due to soy food flavor improvement for human consumption [144]. Overall, conventional breeding has been used to improve protein quality, oil content, and sucrose content in soybean seed.

While soybean does have numerous profitable seed traits, it also contains several unfavorable traits that include trypsin inhibitors (TIs), indigestible carbohydrates, and phytate [145, 146]. There are two trypsin inhibitors found in soybean, the Kunitz and Bowman-Birk trypsin inhibitor, and they are antinutritional factors due to their ability to interfere with protein digestibility and reduce the health of animals that are fed soybean meal containing these proteins [147, 148]. Currently, processors can heat the soybean meal in order to inactivate the trypsin inhibitors, but this step is costly [149]. Due to TIs negatively affecting animal health and increased cost for inactivation, more breeding efforts are being made to develop low-TI soybean lines [150]. Indigestible carbohydrates, raffinose and stachyose, that are found in soybean seed are also a target for soybean breeding since they can cause flatulence and diarrhea when consumed [144]. There has been progress made in lowering these carbohydrates, which include identifying QTLs associated with raffinose and stachyose [144]. Lastly, while phytate is an antinutritional factor found in soybean, there is not a lot of work being done anymore to breed low phytate soybean lines since phytase supplements are an effective, inexpensive way to reduce the phytate found in soybean meal [151].

5.2 Genetic engineering efforts

Genetic engineering involves the process of artificially and intentionally manipulating the DNA of an organism with the purpose of modifying that organism [152]. Some of the methods used to transform plants include *Agrobacterium*, electroporating plant protoplasts, and microparticle bombardment [153]. One relatively new field within genetic engineering is gene editing which involves using clustered regularly interspaced short palindromic repeats or CRISPR/Cas9 system [154]. Genetic engineering works by introducing a gene from one organism into another organism so that it can now express that gene product or by causing frameshifts or deletions to knockout a particular gene in an organism [152]. CRISPR/Cas9 has been gaining a lot of attention due to its promising ability to efficiently and effectively improve agronomic traits in crops [155, 156]. Genetic engineering in soybean was first successfully accomplished in the 1990s [157]. Since this time, genetic engineering has been used frequently in soybean with about 90.7 million hectares of genetically modified/GM soybean being planted in 2014 [157]. Most of this genetic engineering has been done to create Roundup Ready soybean that is resistant to glyphosate herbicides [158]. Roundup Ready soybean is prevalent because it allows growers to spray herbicides to kill any weeds in the field while not killing the soybean [158]. Genetic engineering has been used to additionally improve the protein quality of soybean by altering biosynthetic feedback pathways to increase lysine and by expressing zein proteins from corn to increase sulfur containing amino acids [115, 159]. Besides these examples, genetic engineering has been used to manipulate soybean oil content by increasing oleic acid content and decreasing linolenic acid content and to delay flowering time in soybean [160, 161]. Given the ability of genetic engineering, especially gene editing, to successfully improve qualities of soybean, it will likely be used to improve soybean through removing and/or modifying expression of antinutritional factors. This can be accomplished through genetic engineering by knocking out particular genes responsible for the antinutritional factors preventing them from being expressed.

6. Conclusions

Soybean is an essential crop that is grown globally due to its various and diverse uses. Given its importance, there are many pre-growing practices to prepare the field for the growing season, including tillage, pre-plant fertilization, and monitoring soil pH. Many agronomic aspects must be considered during growing season to ensure successful soybean growth including crop rotations, double cropping, cover crops, irrigation, row spacing, plant density, and post-emergence fertilization. Additionally, integrated pest management involving the use of pesticides, resistant soybean cultivars, and cultural practices are vital to control the numerous pests of soybean. While soybean is highly used in livestock feed due to its high protein content, it's methionine deficiency and presence of antinutritional factors still present problems that need to be solved. Soybean versatility is represented by the many uses in human consumption, biofuels, and other industrial uses. Traditional and conventional breeders have been working to increase protein and oil content, while eliminating antinutritional factors. Genetic engineering and gene editing show promise to help improve soybean by introducing genes to improve protein and oil quality and knocking out genes to remove antinutritional factors.

Acknowledgements

The authors would like to acknowledge Virginia Tech's Open Access Subvention Fund for funding the publishing fee.

Conflict of interest

The authors declare no conflict of interest.

Author details

Zachary Shea*, William M. Singer and Bo Zhang
School of Plant and Environmental Sciences, Virginia Polytechnic Institute and State University, Blacksburg, Virginia, USA

*Address all correspondence to: zacharys@vt.edu

References

[1] International: World Oilseed Production [Internet]. Available from: http://soystats.com/international-world-oilseed-production/

[2] Goldsmith PD. Economics of soybean production, marketing, and utilization. In: Johnson LA, White PJ, Galloway R, editors. Soybeans [Internet]. Urbana, Illinois: AOCS Press; 2008. pp. 117-150. Available from: http://www.sciencedirect.com/science/article/pii/B9781893997646500081

[3] Hartman GL, West ED, Herman TK. Crops that feed the World 2. Soybean—Worldwide production, use, and constraints caused by pathogens and pests. Food Security. 2011;**3**(1):5-17

[4] International: World Soybean Production [Internet]. Available from: http://soystats.com/international-world-soybean-production/

[5] World Soybean Production [Internet]. Soy Meal Info Center; 2018. Available from: https://www.soymeal.org/soy-meal-articles/world-soybean-production/

[6] Hymowitz T, Singh RJ. Taxonomy and speciation. In: Soybeans: Improvement, Production, and Uses. 2nd ed. Madison, WI: ASA, CSSA, SSSA; 1987. (Agronomy)

[7] U.S. Exports: Soy Exports by Customer [Internet]. Available from: http://soystats.com/u-s-exports-soy-exports-by-customersda/

[8] History of Tillage and Tillage Research [Internet]. Available from: https://extension.umn.edu/soil-management-and-health/history-tillage-and-tillage-research

[9] NC State Extension. North Carolina Soybean Production Guide. North Carolina State University; 2018

[10] No-Till Research [Internet]. Available from: http://milan.tennessee.edu/research/notill.asp

[11] Al-Kaisi M, Licht M. Is tillage needed for your soybean crop? [Internet]. Intergrated Crop Management. Available from: https://crops.extension.iastate.edu/encyclopedia/tillage-needed-your-soybean-crop

[12] Yusuf RI, Siemens JC, Bullock DG. Growth analysis of soybean under no-tillage and conventional tillage systems. Agronomy Journal. 1999;**91**(6):928-933

[13] Rathke G-W, Wienhold BJ, Wilhelm WW, Diepenbrock W. Tillage and rotation effect on corn–soybean energy balances in eastern Nebraska. Soil and Tillage Research. 2007;**97**(1):60-70

[14] Popp MP, Keisling TC, McNew RW, Oliver LR, Dillon CR, Wallace DM. Planting date, cultivar, and tillage system effects on dryland soybean production. Agronomy Journal. 2002;**94**(1):81-88

[15] Vetsch JA, Randall GW, Lamb JA. Corn and soybean production as affected by tillage systems. Agronomy Journal. 2007;**99**(4):952-959

[16] Pedersen P, Lauer JG. Corn and soybean response to rotation sequence, row spacing, and tillage system. Agronomy Journal. 2003;**95**(4):965-971

[17] Fernandez-Cornejo J, Hallahan C, Nehring R, Wechsler S, Grube A. Conservation tillage, herbicide use, and genetically engineered crops in the United States: The case of soybeans. AgBioForum. 2012;**15**(3):231-241

[18] Ferreira MC, Andrade DS, Chueire LMO, Takemura SM, Hungria M.

Tillage method and crop rotation effects on the population sizes and diversity of bradyrhizobia nodulating soybean. Soil Biology and Biochemistry. 2000;**32**(5):627-637

[19] Okada H, Harada H. Effects of tillage and fertilizer on nematode communities in a Japanese soybean field. Applied Soil Ecology. 2007;**35**(3):582-598

[20] Carter MR, Gregorich EG, editors. Soil Sampling and Methods of Analysis. 2nd ed. Boca Raton, FL: CRC Press; 2007

[21] Ferguson RB. EC06-155 nutrient management for agronomic crops in Nebraska. In: Historical Materials from University of Nebraska-Lincoln Extension. 2006. pp. 121-126

[22] Orf JH, Diers BW, Boerma HR. Managing inputs for peak production. In: Soybeans: Improvement, Production, and Uses. 3rd ed. Madison, WI: ASA, CSSA, SSSA; 2004. pp. 451-525. (Agronomy)

[23] Bender RR, Haegele JW, Below FE. Nutrient uptake, partitioning, and remobilization in modern soybean varieties. Agronomy Journal. 2015;**107**(2):563-573

[24] Varco JJ. Nutrition and fertility requirements. In: Soybean Production in the Mid-South. Boca Raton, FL: CRC Press; 1999. pp. 53-70

[25] Zhou X-J, Liang Y, Chen H, Shen S-H, Jing Y-X. Effects of rhizobia inoculation and nitrogen fertilization on photosynthetic physiology of soybean. Photosynthetica. 2006;**44**(4):530-535

[26] Brann DE, Abaye AO, Azenegashe O, Peterson PR, Paul R, Chalmers DR, Whitt DL, Chappell GF, et al. Agronomy Handbook. 2009. Available from: https://vtechworks.lib.vt.edu/handle/10919/48840

[27] Troeh FR, Thompson LM. Soils and Soil Fertility. 6th ed. Ames, Iowa: Blackwell Publishing Professional; 2005

[28] Carrera CS, Reynoso CM, Funes GJ, Martínez MJ, Dardanelli J, Resnik SL. Amino acid composition of soybean seeds as affected by climatic variables. Brazilian Agricultural Research. 2011;**46**(12):1579-1587

[29] Goldflus F, Ceccantini M, Santos W. Amino acid content of soybean samples collected in different Brazilian states: Harvest 2003/2004. Brazilian Journal of Poultry Science. 2006;**8**(2):105-111

[30] Wolf RB, Cavins JF, Kleiman R, Black LT. Effect of temperature on soybean seed constituents: Oil, protein, moisture, fatty acids, amino acids and sugars. Journal of the American Oil Chemists' Society. 1982;**59**(5):230-232

[31] Krishnan HB, Bennett JO, Kim W-S, Krishnan AH, Mawhinney TP. Nitrogen lowers the sulfur amino acid content of soybean (*Glycine max* [L.] Merr.) by regulating the accumulation of Bowman–Birk protease inhibitor. Journal of Agricultural and Food Chemistry. 2005;**53**(16):6347-6354

[32] Kaur G, Serson WR, Orlowski JM, McCoy JM, Golden BR, Bellaloui N. Nitrogen sources and rates affect soybean seed composition in Mississippi. Agronomy. 2017;**7**(4):77

[33] Raboy V, Dickinson DB. Effect of phosphorus and zinc nutrition on soybean seed phytic acid and zinc. Plant Physiology. 1984;**75**(4):1094-1098

[34] Kapoora AC, Gupta YP. Changes in proteins and amino acids in developing soybean seed and effect of phosphorus nutrition. Journal of the Science of Food and Agriculture. 1977;**28**(2):113-120

[35] Vyn TJ, Yin X, Bruulsema TW, Jackson C-JC, Rajcan I, Brouder SM.

Potassium fertilization effects on isoflavone concentrations in soybean [*Glycine max* (L.) Merr.]. Journal of Agricultural and Food Chemistry. 2002;**50**(12):3501-3506

[36] Myers SW, Gratton C, Wolkowski RP, Hogg DB, Wedberg JL. Effect of soil potassium availability on soybean aphid (Hemiptera: Aphididae) population dynamics and soybean yield. Journal of Economic Entomology. 2005;**98**(1):113-120

[37] Walter AJ, DiFonzo CD. Soil potassium deficiency affects soybean phloem nitrogen and soybean aphid populations. Environmental Entomology. 2007;**36**(1):26-33

[38] Devi KN, Singh LNK, Sumarjit Singh M, Basanta Singh S, Khamba Singh K. Influence of sulphur and boron fertilization on yield, quality, nutrient uptake and economics of soybean (*Glycine max*) under upland conditions. The Journal of Agricultural Science. 2012;**4**(4):1

[39] Gayler KR, Sykes GE. Effects of nutritional stress on the storage proteins of soybeans. Plant Physiology. 1985;**78**(3):582-585

[40] Wilhelm WW, Wortmann CS. Tillage and rotation interactions for corn and soybean grain yield as affected by precipitation and air temperature. Agronomy Journal. 2004;**96**(2): 425-432

[41] Karlen DL, Hurley EG, Andrews SS, Cambardella CA, Meek DW, Duffy MD, et al. Crop rotation effects on soil quality at three northern corn/soybean belt locations. Agronomy Journal. 2006;**98**(3):484-495

[42] Balota EL, Colozzi-Filho A, Andrade DS, Dick RP. Microbial biomass in soils under different tillage and crop rotation systems. Biology and Fertility of Soils. 2003;**38**(1):15-20

[43] Hoeft RG, Nafziger ED, Johnson RR, Aldrich SR. Modern Corn and Soybean Production. 1st ed. Champaign, IL: MCSP Publishing; 2000

[44] Mirsky SB, Curran WS, Mortenseny DM, Ryany MR, Shumway DL. Timing of cover-crop management effects on weed suppression in no-till planted soybean using a roller-crimper. Weed Science. 2011;**59**(3):380-389

[45] Williams SM, Weil RR. Crop cover root channels may alleviate soil compaction effects on soybean crop. Soil Science Society of America Journal. 2004;**68**(4):1403-1409

[46] Davis AS. Cover-crop roller–crimper contributes to weed management in no-till soybean. Weed Science. 2010;**58**(3):300-309

[47] Reddy KN. Effects of cereal and legume cover crop residues on weeds, yield, and net return in soybean (*Glycine max*). Weed Technology. 2001;**15**(4):660-668

[48] Shapiro BI, Brorsen BW, Doster DH. Adoption of double-cropping soybeans and wheat. Journal of Agricultural and Applied Economics. 1992;**24**(2):33-40

[49] Kelley KW. Double-cropping winter wheat and soybean improves net returns in the eastern great plains. Crop Management. 2003;**2**(1)

[50] Jones BP, Holshouser DL, Alley MM, Roygard JKF, Anderson-Cook CM. Double-crop soybean leaf area and yield responses to mid-atlantic soils and cropping systems. Agronomy Journal. 2003;**95**(2):436-445

[51] Egli DB, Bruening WP. Potential of early-maturing soybean cultivars in late plantings. Agronomy Journal. 2000;**92**(3):532-537

[52] Hayder G, Mumtaz SS, Khan A, Khan S. Maize and soybean intercropping under various levels of soybean seed rates. Asian Journal of Plant Sciences. 2003;**2**(3):339-341

[53] Muoneke CO, Ogwuche MAO, Kalu BA. Effect of maize planting density on the performance of maize/soybean intercropping system in a guinea savannah agroecosystem. African Journal of Agricultural Research. 2008;**2**(12):667-677

[54] Li L, Sun J, Zhang F, Li X, Rengel Z, Yang S. Wheat/maize or wheat/soybean strip intercropping: II. Recovery or compensation of maize and soybean after wheat harvesting. Field Crops Research. 2001;**71**(3):173-181

[55] Zhang F, Li L. Using competitive and facilitative interactions in intercropping systems enhances crop productivity and nutrient-use efficiency. Plant and Soil. 2003;**248**(1):305-312

[56] Li X, Mu Y, Cheng Y, Liu X, Nian H. Effects of intercropping sugarcane and soybean on growth, rhizosphere soil microbes, nitrogen and phosphorus availability. Acta Physiologiae Plantarum. 2013;**35**(4):1113-1119

[57] Hock SM, Knezevic SZ, Martin AR, Lindquist JL. Soybean row spacing and weed emergence time influence weed competitiveness and competitive indices. Weed Science. 2006;**54**(1):38-46

[58] Knezevic SZ, Evans SP, Mainz M. Row spacing influences the critical timing for weed removal in soybean (*Glycine max*). Weed Technology. 2003;**17**(4):666-673

[59] Arce GD, Pedersen P, Hartzler RG. Soybean seeding rate effects on weed management. Weed Technology. 2009;**23**(1):17-22

[60] Holshouser DL, Whittaker JP. Plant population and row-spacing effects on early soybean production systems in the mid-Atlantic USA. Agronomy Journal. 2002;**94**(3):603-611

[61] Cox WJ, Cherney JH. Growth and yield responses of soybean to row spacing and seeding rate. Agronomy Journal. 2011;**103**(1):123-128

[62] De Bruin JL, Pedersen P. Effect of row spacing and seeding rate on soybean yield. Agronomy Journal. 2008;**100**(3):704-710

[63] Lambert DM, Lowenberg-DeBoer J. Economic analysis of row spacing for corn and soybean. Agronomy Journal. 2003;**95**(3):564-573

[64] Specht JE, Chase K, Macrander M, Graef GL, Chung J, Markwell JP, et al. Soybean response to water. Crop Science. 2001;**41**(2):493-509

[65] Heatherly LG. Soybean irrigation. In: Soybean Production in the Mid-South. Boca Raton, FL: CRC Press; 1999. pp. 119-142

[66] Specht JE, Hume DJ, Kumudini SV. Soybean yield potential—A genetic and physiological perspective. Crop Science. 1999;**39**(6):1560-1570

[67] Arora VK, Singh CB, Sidhu AS, Thind SS. Irrigation, tillage and mulching effects on soybean yield and water productivity in relation to soil texture. Agricultural Water Management. 2011;**98**(4):563-568

[68] Karam F, Masaad R, Sfeir T, Mounzer O, Rouphael Y. Evapotranspiration and seed yield of field grown soybean under deficit irrigation conditions. Agricultural Water Management. 2005;**75**(3):226-244

[69] Nijbroek R, Hoogenboom G, Jones JW. Optimizing irrigation

management for a spatially variable soybean field. Agricultural Systems. 2003;**76**(1):359-377

[70] Mallarino AP, Haq MU, Wittry D, Bermudez M. Variation in soybean response to early season foliar fertilization among and within fields. Agronomy Journal. 2001;**93**(6):1220-1226

[71] Freeborn JR, Holshouser DL, Alley MM, Powell NL, Orcutt DM. Soybean yield response to reproductive stage soil-applied nitrogen and foliar-applied boron. Agronomy Journal. 2001;**93**(6):1200-1209

[72] Salvagiotti F, Specht JE, Cassman KG, Walters DT, Weiss A, Dobermann A. Growth and nitrogen fixation in high-yielding soybean: Impact of nitrogen fertilization. Agronomy Journal. 2009;**101**(4):958-970

[73] Haq MU, Mallarino AP. Soybean yield and nutrient composition as affected by early season foliar fertilization. Agronomy Journal. 2000;**92**(1):16-24

[74] Haq MU, Mallarino AP. Response of soybean grain oil and protein concentrations to foliar and soil fertilization. Agronomy Journal. 2005;**97**(3):910-918

[75] Popović V, Glamočlija Đ, Sikora V, Đekić V, Červenski J, Simić D, et al. Genotypic specificity of soybean [*Glycine max* (L) Merr.] under conditions of foliar fertilization. Romanian Agricultural Research. 2013;(30):259-270

[76] Xiaoming Z, Qiong L. A brief introduction of main diseases and insect pests in soybean production in the global top five soybean producing countries. Plant Diseases & Pests. 2018;**9**(1):17-21

[77] Lahiri S, Reisig DD. Ecology and management of Kudzu Bug (Hemiptera: Plataspidae) in southeastern soybeans. Journal of Integrated Pest Management. 2016;**7**(1). Available from: https://academic.oup.com/jipm/article/7/1/14/2658149

[78] Mexican Bean Beetle—Epilachna varivestis Mulsant [Internet]. Available from: http://entnemdept.ufl.edu/creatures/veg/bean/mexican_bean_beetle.htm

[79] USDA APHIS | Japanese Beetle [Internet]. Available from: https://www.aphis.usda.gov/aphis/ourfocus/planthealth/plant-pest-and-disease-programs/pests-and-diseases/japanese-beetle/japanese-beetle

[80] Managing Spider Mite on Soybean [Internet]. Available from: https://extension.umn.edu/soybean-pest-management/managing-spider-mite-soybean

[81] Leskey T, Hamilton G, Nielsen A, Polk D, Rodriquez-Saona C, Christopher J, et al. Pest Status of the Brown Marmorated Stink Bug. Halyomorpha Halys in the USA: Outlooks Pest Manag; 2012

[82] Sustainable Solutions to Pest Problems [Internet]. Integrated Pest Management. 2019. Available from: https://www.sccgov.org/sites/ipm/about/Pages/benefits-of-ipm.aspx

[83] Norris R. Integrated pest management. In: Encyclopedia of Biological Invasions [Internet]. 2004. Available from: https://books.google.com/books?hl=en&lr=&id=qgQmDQAAQBAJ&oi=fnd&pg=PA353&dq=integrated+pest+management&ots=T9pG1zlok0&sig=dl87ucNN0OuYWpZ9AhVwMaf6AHE#v=onepage&q=integrated%20pest%20management&f=false

[84] Motaphale AA, Bhosle BB. Validation of integrated pest management modules on defoliating insect pests of soybean. Agricultural Science Digest. 2016;**36**(4):303-306

[85] Hokkanen HMT. Trap cropping in pest management. Annual Review of Entomology. 1991;**36**:119-138

[86] Soybean Insect Management | Extending Agricultural Information [Internet]. Mississippi Crop Situation. 2017. Available from: https://www.mississippi-crops.com/insect-control-guide/soybean-insect-management/

[87] Vivian R, Reis A, Kálnay PA, Vargas L, Ferreira ACC, Mariani F. Weed Management in Soybean—Issues and Practices. Soybean—Pest Resist [Internet]. 2013. Available from: https://www.intechopen.com/books/soybean-pest-resistance/weed-management-in-soybean-issues-and-practices

[88] Oerke EC, Dehne HW. Safeguarding production—Losses in major crops and the role of crop protection. Available from: https://www.sciencedirect.com/science/article/pii/S0261219403002540

[89] Hartzler B. Managing weeds to protect crop yields. In: Integrated Crop Management [Internet]. Iowa State University Extension and Outreach. Available from: https://crops.extension.iastate.edu/encyclopedia/managing-weeds-protect-crop-yields

[90] Prins A. Identifying troublesome broadleaf weeds in soybeans [Internet]. IL Soy Advisor. 2018. Available from: https://www.ilsoyadvisor.com/on-farm/ilsoyadvisor/identifying-troublesome-broadleaf-weeds-soybeans

[91] Identifying Weeds in Field Crops. Integrated Pest Management. 2015

[92] Cultural Weed Management [Internet]. NC State Extension. 2013. Available from: https://soybeans.ces.ncsu.edu/cultural-weed-management/

[93] Sprague C. Herbicide Classification Guide. Soy Checkoff; 2016

[94] Postemergence Herbicide Applications in Soybean [Internet]. Available from: https://www.roundupreadyplus.com/resourcecenter/postemergence_herbicide_applications_in_soybean/

[95] Hemmati P, Zafari D, Mahmoodi SB, Hashemi M, Gholamhoseini M, Dolatabadian A, et al. Histopathology of charcoal rot disease (*Macrophomina phaseolina*) in resistant and susceptible cultivars of soybean. Rhizosphere. 2018;7:27-34

[96] Malvick D. Charcoal Rot on Soybean [Internet]. University of Minnesota Extension. 2018. Available from: https://extension.umn.edu/pest-management/charcoal-rot-soybean

[97] GieslerLJ,BroderickKC.Management of Phytophthora Root and Stem Rot of Soybean. Univ Neb-Linc Ext; 2017. p. 5

[98] Giesler LJ. Bacterial Diseases of Soybean. Univ Neb-Linc Ext; 2011. p. 2

[99] Malvick D. Soybean Rust [Internet]. 2018. Available from: https://extension.umn.edu/pest-management/soybean-rust

[100] Malvick D. Anthracnose on Soybean [Internet]. 2018. Available from: https://extension.umn.edu/pest-management/anthracnose-soybean

[101] Malvick D. Septoria Brown Spot [Internet]. 2018. Available from: https://extension.umn.edu/pest-management/septoria-brown-spot

[102] Giesler LJ. Soybean Cyst Nematode. Univ Neb-Linc Ext [Internet]; 2015. Available from: https://cropwatch.unl.edu/plantdisease/soybean/soybean-cyst-nematode

[103] Athow KL. Soybean pest management. Journal of the American Oil Chemists' Society. 1981;**58**(3):130-135

[104] Nagasubramanian K, Jones S, Sarkar S, Singh AK, Singh A, Ganapathysubramanian B. Hyperspectral band selection using genetic algorithm and support vector machines for early identification of charcoal rot disease in soybean stems. Plant Methods. 2018;**14**(1):86

[105] Brucker E, Niblack T, Kopisch-Obuch F, Diers B. The effect of rhg1 on reproduction of heterodera glycines in the field and greenhouse and associated effects on agronomic traits. Crop Science. 2005;**45**:1721-1727

[106] Guo W, Zhang F, Bao A, You Q, Li Z, Chen J, et al. The soybean Rhg1 amino acid transporter gene alters glutamate homeostasis and jasmonic acid-induced resistance to soybean cyst nematode. Molecular Plant Pathology. 2019;**20**(2):270-286

[107] Niblack TL. Soybean cyst nematode management reconsidered. Plant Disease. 2005;**89**(10):1020-1026

[108] Heuze V, Tran G, Kaushik S. Soybean Meal. Feedipedia [Internet]. Update by hand. Available from: https://www.feedipedia.org/node/674

[109] Soybean Meal [Internet]. United Soybean Board. Available from: https://unitedsoybean.org/topics/soybean-meal/

[110] Buttery PJ, D'Mello JPF. Amino acid metabolism in farm animals: An overview. In: Amino Acids in Farm Animal Nutrition. Wallingford, United Kingdom: CAB International; 1994. pp. 1-10

[111] Kuiken KA, Lyman M. Essential amino acid composition of soy bean meals prepared from twenty strains of soy beans. The Journal of Biological Chemistry. 1949;**177**(1):29-36

[112] Wolawek-Potocka I, Bah MM, Korzekwa A, Piskula MK, Wiczkowski W, Depta A, et al. Soybean-derived phytoestrogens regulate prostaglandin secretion in endometrium during cattle estrous cycle and early pregnancy. Experimental Biology and Medicine. 2005;**230**(3):189-199

[113] McNiven MA, Duynisveld J, Charmley E, Mitchell A. Processing of soybean affects meat fatty acid composition and lipid peroxidation in beef cattle. Animal Feed Science and Technology. 2004;**116**(3):175-184

[114] Lusas EW. Soybean processing and utilization. In: Soybeans: Improvement, Production, and Uses. 3rd ed. Madison, WI: ASA, CSSA, SSSA; 2004. pp. 949-1036. (Agronomy)

[115] Liu K. Soybeans: Chemistry, Technology, and Utilization. New York, NY: Chapman & Hall; 1997

[116] Berry TH, Becker DE, Rasmussen OG, Jensen AH, Norton HW. The limiting amino acids in soybean protein. Journal of Animal Science. 1962;**21**(3):558-561

[117] Boisen S, Hvelplund T, Weisbjerg MR. Ideal amino acid profiles as a basis for feed protein evaluation. Livestock Production Science. 2000;**64**(2):239-251

[118] Fernandez SR, Aoyagi S, Han Y, Parsons CM, Baker DH. Limiting order of amino acids in corn and soybean meal for growth of the chick. Poultry Science. 1994;**73**(12):1887-1896

[119] Bunchasak C. Role of dietary methionine in poultry production. The Journal of Poultry Science. 2009;**46**(3):169-179

[120] Han Y, Suzuki H, Parsons CM, Baker DH. Amino acid fortification of a low-protein corn and soybean meal diet for chicks. Poultry Science. 1992;**71**(7):1168-1178

[121] Managing Nutrient and Pathogens from Animal Agriculture. Ithaca, NY: Natural Resource, Agriculture, and Enegineering Service; 2000

[122] Control Feed Costs with Amino Acids [Internet]. Drovers. Available from: https://www.drovers.com/article/control-feed-costs-amino-acids

[123] Vagadia BH, Vanga SK, Raghavan V. Inactivation methods of soybean trypsin inhibitor—A review. Trends in Food Science and Technology. 2017;**64**:115-125

[124] Golbitz P. Traditional soyfoods: Processing and products. Journal of Nutrition. 1995;**125**(suppl_3):570S-572S

[125] Birt DF, Hendrich S, Alekel DL, Anthony M. Soybean and the prevention of chronic human disease. In: Soybeans: Improvement, Production, and Uses. 3rd ed. Madison, WI: ASA, CSSA, SSSA; 2004. pp. 1047-1103. (Agronomy)

[126] Messina MJ. Soyfoods: Their role in disease prevention and treatment. In: Soybeans: Chemistry, Technology, and Utilization. New York, NY: Chapman & Hall; 1997

[127] Messina M. Soy foods, isoflavones, and the health of postmenopausal women. American Journal of Clinical Nutrition. 2014;**100**(suppl_1):423S-430S

[128] Messina M. A brief historical overview of the past two decades of soy and isoflavone research. The Journal of Nutrition. 2010;**140**(7):1350S-1354S

[129] Xiao Y, Zhang S, Tong H, Shi S. Comprehensive evaluation of the role of soy and isoflavone supplementation in humans and animals over the past two decades. Phytotherapy Research. 2018;**32**(3):384-394

[130] Zhong X, Ge J, Chen S, Xiong Y, Ma S, Chen Q. Association between dietary isoflavones in soy and legumes and endometrial cancer: A systematic review and meta-analysis. Journal of the Academy of Nutrition and Dietetics. 2018;**118**(4):637-651

[131] Biodiesel [Internet]. United Soybean Board. Available from: https://www.unitedsoybean.org/media-center/issue-briefs/biodiesel/

[132] Cavalett O, Ortega E. Integrated environmental assessment of biodiesel production from soybean in Brazil. Journal of Cleaner Production. 2010;**18**(1):55-70

[133] Özener O, Yüksek L, Ergenç AT, Özkan M. Effects of soybean biodiesel on a DI diesel engine performance, emission and combustion characteristics. Fuel. 2014;**115**:875-883

[134] Pradhan A, Shrestha DS, McAloon A, Yee W, Haas M, Duffield JA, et al. Energy life-cycle assessment of soybean biodiesel. Agricultural Economic Report. 2009;**845**:31

[135] Sudarić A, Kočar MM, Duvnjak T, Zdunić Z, Kulundžić AM. Improving seed quality of soybean suitable for growing in Europe. Soybean Hum Consum Anim Feed [Internet]. 2019. Available from: https://www.intechopen.com/online-first/improving-seed-quality-of-soybean-suitable-for-growing-in-europe

[136] Zarkadas C, Voldeng H, Yu Z, Choi V. Assessment of the protein quality of nine northern adapted yellow and brown seed coated soybean cultivars by amino acid analysis. Journal of Agricultural and Food Chemistry. 1999;**47**(12):5009-5018

[137] Kurasch AK, Hahn V, Leiser WL, Starck N, Würschum T. Phenotypic analysis of major agronomic traits in 1008 RILs from a diallel of early european soybean varieties. Crop Science. 2017;**57**(2):726-738

[138] Wilcox JR, Shibles RM. Interrelationships among seed quality attributes in soybean. Crop Science. 2001;**41**(1):11-14

[139] Vaughn JN, Nelson RL, Song Q, Cregan PB, Li Z. The genetic architecture of seed composition in soybean is refined by genome-wide association scans across multiple populations. G3: Genes, Genomes, Genetics. 2014;**4**(11):2283-2294

[140] Gupta PK, Kulwal PL, Mir RR. QTL mapping: Methodology and applications in cereal breeding. In: Cereal Genomics II [Internet]. New York City, New York: Springer; 2013. pp. 275-318. Available from: https://link-springer-com.ezproxy.lib.vt.edu/chapter/10.1007/978-94-007-6401-9_11

[141] Fehr WR. Breeding for modified fatty acid composition in soybean. Crop Science. 2007;**47**:S-72-S-87

[142] Lopez S, Bermudez B, Pacheco YM, Ortega A, Varela LM, Abia R, et al. Chapter 154—Oleic acid: The main component of olive oil on postprandial metabolic processes. In: Preedy VR, Watson RR, editors. Olives and Olive Oil in Health and Disease Prevention [Internet]. San Diego: Academic Press; 2010. pp. 1385-1393. Available from: http://www.sciencedirect.com/science/article/pii/B9780123744203001546

[143] Hou A, Chen P, Alloatti J, Li D, Mozzoni L, Zhang B, et al. Genetic variability of seed sugar content in worldwide soybean germplasm collections. Crop Science. 2009;**49**(3):903-912

[144] Wang Y, Chen P, Zhang B. Quantitative trait loci analysis of soluble sugar contents in soybean. Plant Breeding. 2014;**133**(4):493-498

[145] El-Shemy H, Abdel-Rahim E, Shaban O, Ragab A, Carnovale E, Fujita K. Comparison of nutritional and antinutritional factors in soybean and fababean seeds with or without cortex. Soil Science & Plant Nutrition. 2000;**46**(2):515-524

[146] Refstie S, Sahlström S, Bråthen E, Baeverfjord G, Krogedal P. Lactic acid fermentation eliminates indigestible carbohydrates and antinutritional factors in soybean meal for Atlantic salmon (*Salmo salar*). Aquaculture. 2005;**246**(1):331-345

[147] Barrows FT, Stone DAJ, Hardy RW. The effects of extrusion conditions on the nutritional value of soybean meal for rainbow trout (*Oncorhynchus mykiss*). Aquaculture. 2007;**265**(1):244-252

[148] Becker-Ritt AB, Mulinari F, Vasconcelos IM, Carlini CR. Antinutritional and/or toxic factors in soybean (*Glycine max* (L) Merril) seeds: comparison of different cultivars adapted to the southern region of Brazil. Journal of the Science of Food and Agriculture. 2004;**84**(3):263-270

[149] Snowdon M. Full-fat soybeans for livestock—Agriculture [Internet]. Agriculture, Aquaculture, and Fisheries. 2012. Available from: https://www2.gnb.ca/content/gnb/en/departments/10/agriculture/content/crops/field_crops/Soybeans.html

[150] Hymowitz T. Genetics and breeding of soybeans lacking the kunitz trypsin inhibitor. In: Friedman M, editor. Nutritional and Toxicological Significance of Enzyme Inhibitors in Foods [Internet], Advances in Experimental Medicine and Biology. Boston, MA: Springer US; 1986. pp. 291-298. DOI: 10.1007/978-1-4757-0022-0_18

[151] Lei XG, Porres JM. Phytase: An enzyme to improve soybean nutrition. In: Soybean and Nutrition [Internet]. 2011. Available from: https://www.intechopen.com/books/

soybean-and-nutrition/phytase-an-enzyme-to-improve-soybean-nutrition

[152] Augsutyn A. Genetic engineering | Definition, Process, & Uses [Internet]. Encyclopedia Britannica. 2019. Available from: https://www.britannica.com/science/genetic-engineering

[153] National Academies of Science. Methods and mechanisms for genetic manipulation of plants, animals, and microorganisms. In: Safety of Genetically Engineered Foods: Approachers to Assessing Unintended Health Effects [Internet]. National Academies Press; 2004. Available from: https://www.ncbi.nlm.nih.gov/books/NBK215771/

[154] What are genome editing and CRISPR-Cas9?—Genetics Home Reference—NIH [Internet]. Genetics Home Reference. 2020. Available from: https://ghr.nlm.nih.gov/primer/genomicresearch/genomeediting

[155] Arora L, Narula A. Gene Editing and Crop Improvement Using CRISPR-Cas9 System [Internet]. Frontiersin. 2017. Available from: https://www.frontiersin.org/articles/10.3389/fpls.2017.01932/full

[156] Wang T, Zhang H, Zhu H. CRISPR technology is revolutionizing the improvement of tomato and other fruit crops. Horticulture Research [Internet]. 2019. Available from: https://www.nature.com/articles/s41438-019-0159-x

[157] Biotech Crop Highlights in 2018 | ISAAA.org [Internet]. Available from: http://www.isaaa.org/resources/publications/pocketk/16/

[158] Roundup Ready Soybeans—SourceWatch [Internet]. Available from: https://www.sourcewatch.org/index.php/Roundup_Ready_Soybeans

[159] Falco SC, Locke M, Guida T, Sanders C, Ward RT, Webber P. Transgenic canola and soybean seeds with increased lysine. Nature Biotechnology. 1995;**13**:577-582

[160] Clemente T, Cahoon E. Soybean oil: Genetic approaches for modification of functionality and total content. Plant Physiology [Internet]. 2009. Available from: https://www.ncbi.nlm.nih.gov/pmc/articles/PMC2773065/

[161] Cai Y, Chen L, Guo C, Sun S, Wu C, Jiang B, et al. CRISPR/Cas9-mediated targeted mutagenesis of GmFT2a delays flowering time in soya bean. Plant Biotechnology Journal. 2017;**16**(1):176-185

Chapter 4

Food Grade Soybean Breeding, Current Status and Future Directions

Souframanien Jegadeesan and Kangfu Yu

Abstract

Soybeans possess average 20% oil and 40% protein content and are a major source of protein and fatty acids in human and animal nutrition. Soybean cultivars are classified as commodity type, which are used for edible or industrial oil and animal feed, and food-type, which are used for human consumption in fermented foods and non-fermented foods. Major breeding targets for food grade soybeans are high protein and sucrose content. Developing cultivars with desired seed size and appearance depends on the type of soyfood for which the soybeans are destined. Seed with high protein content (>45%), low oil content, high sucrose, and low oligosaccharide content are suitable for making soymilk and tofu. For soyfood such as natto, soybean seed with a high content of carbohydrates are preferred. Since, molecular markers linked to the target food traits have been developed, transfer of the food grade traits among soybean varieties is possible through marker-assisted selection (MAS) to track the target gene/QTLs. Introgression of wild soybean alleles through genomics assisted breeding (e.g., GWAS, haplotype blocks, NIL, etc.), high-throughput phenotyping, mutagenesis and genome engineering/editing would improve protein without yield drag, pleiotropic effects, and background/allelic effects in breeding food grade soybean.

Keywords: soybean, soyfood, selection, mar-assisted selection

1. Introduction

Soybean [*Glycine max* (L.) Merr.] is one of the most important crops in the world. Soybean seeds are distinguished from other legume grains especially by their high protein content. It is a major source of vegetable protein and oil in human food industries because of its nutritive and health benefits [1]. Owing to its health benefits, soy food has been part of the Asian diet for several centuries [2]. Soyfood has nutritional qualities that reduce human blood serum cholesterol levels and lower the risk of cardiovascular diseases [3]. Soy food is a natural source of isoflavones (daidzein, genistein, and glycitein) and a good source of calcium [4].

Soybeans typically possess protein and oil contents of approximately 40 and 20%, respectively. This composition gives the possibility for a broad variety of applications such as feed, biodiesel, edible oils, and other food products. Commercially, soybean can be categorized as (i) commodity type, mainly used for oil and animal feed, and (ii) food type soybean, mainly used for human consumption. One important application of soybean is its use in animal feed. Approximately 85%

of the soybean is crushed into oil and meal. The greatest part of the meal is used as protein source in animal feed, Approximately 6% of the produced soybean is directly used for food purposes [5]. When soybean is exclusively used for consumption by the farm animals, nutritionally, it should have at least 36% protein to meet part of their daily requirement for protein. The food-type soybean, on the other hand, contains protein higher than feed-grade soybean; it usually ranges between 40 and 45%. It helps in making better quality soy-based food products. Usually, food-grade soybeans are used for production of soy milk, tofu or soy paneer, soy sprouts, and other soy-based food items.

Globally soybean meal accounts for 63% of all the protein sources in animal feed. Next to the soybean meal, soybean oil obtained after crushing is refined and used as cooking oils and salad oils, and they can also find their use in mayonnaise and butter substitute. Soybean oil is refined and lecithin is obtained as a main by-product. Lecithins are found in many products such as in chocolate, margarine, and emulsifying agents [6]. Furthermore, the unraveling of the health and nutritional benefits of soybean contributed to increased interest for soybeans destined for food production, the so-called food-grade soybeans. Hence, the breeding of cultivars with enhanced food-grade traits is gaining great importance [7]. Soy foods available in the market can be divided in traditional and non-traditional soy foods. Traditional soy foods includes edamame (green vegetable soybeans), soy milk (a drink produced by soaking and grinding soybeans, boiling the mixture, and filtering out remaining particulates), tofu (soybean curd), tempeh (made of whole cooked soybeans), miso (fermented soup-based paste), soy sauce, okara, natto (fermented whole soybeans), and soy sprouts. They originated from Asia where soybean has been grown for centuries before its introduction to the rest of the world. Non-traditional foods are, for instance, soy yoghurt, soy cheese, pudding, snacks, etc. [6]. Soy food available in the market is classified into two groups based on the soybean seed size. Soy foods made of large seeds (>20 g/100 seeds) include tofu, edamame, miso, and soy milk. Soy foods made of small seeds (<12 g/100 seeds) include natto, soy sauce, tempeh, and bean sprouts.

Soybean seed needs to satisfy specific physical and chemical requirements for soy food production. In addition to seed size, visual appearance such as uniformity of seed size and shape with light-colored hilum and yellow seed coat without physical damage such as mottling, splits, shriveling, purple stain, and insect damage are the main consideration of food-grade soybeans [8]. It is reported that the seed size uniformity affected water absorption and the quality of the final soy product [9]. Shrunken or discolored seeds were undesirable due to the consumer requirements [10]. Stone seeds that do not absorb water during soaking cause serious problems for food processing as it affects the texture and consistency of the soy products particularly for fermented soy food such as natto [11]. Seeds with harder texture have higher calcium content and absorb less water [12]. Seed's hardness could be estimated by the seed swell ratio that was related to seed weight or water volume change before and after soaking [13]. High water absorption was required to obtain soft steamed seeds [14]. Thus, soybean seeds with rapid and high water uptake are preferred by processors in order to provide more products per unit of time [9]. However, seed composition requirements vary according to the type of soy food. Soybean seeds with high protein content (>45%), low oil content, high sucrose content, and low oligosaccharides (raffinose and stachyose) content are suitable for making tofu. For soy food such as natto made through short fermentation process, soybean seed with a high carbohydrate content are preferred for the purpose of getting a quick conversion to simple sugars [10].

2. Anti-nutritional factors

Like other plants, soybean too synthesizes a range of secondary metabolites for their adaptation and self-protection, called anti-nutritional factor (ANF). From nutritional point of view, the ANF is considered as harmful and toxic, as it interferes with normal growth, reproduction, and health. Therefore, soybean, in general, and food-grade soybeans, in particular, are expected to be free from anti-nutritional factors or allergens that may cause harm to the consumers. Soybean contains several anti-nutritional factors among which trypsin inhibitor (TI) and phytase are the most important ones. Kunitz trypsin inhibitor (KTI), which constitutes more than 80% of the total TI, is primarily responsible for improper digestibility of soybean, if consumed unprocessed. Although KTI is heat labile, yet heat inactivation process is neither fully effective nor economic. Genetic elimination of KTI is the most effective way of making soybean free from KTI [15].

2.1 Phytase

Soybean seeds contain phosphorus in the form of inositol hexaphosphate commonly known as phytate [16]. Besides sequestering inorganic phosphate, phytate may also chelate divalent cations such as Fe, thereby decreasing their availability. It also contributes toward water pollution by eutrophication as the phosphate-rich waste discharges into water bodies. So, reduction of seed phytase can enhance mineral and protein bioavailability in soybean. Microarray-based gene expression profiling of phytic acid biosynthesis pathway indicated stepwise regulation of eight genes, viz. myo-inositol-3-phosphatesynthaes (MIPS), inositol phosphate kinase (IPK1–4), etc. Gene silencing constructs were used to silence *GmIKP1* and *GmIPK2* through seed-specific vicilin/conglycinin promoters [17]. Similarly, mutation breeding through gamma irradiation has also been attempted to develop plants with reduced or zero phytase.

3. Cadmium content

Vast areas of agricultural soils are contaminated with Cadmium (Cd) through the use of super phosphate fertilizers, sewage sludge, and inputs from the mining and smelting industries [18]. Cadmium (Cd) is a highly toxic element for human beings because of its extremely long biological half-life. Soybeans grown in cadmium-contaminated soil take up cadmium by roots and translocate into aerial organs, where it affects photosynthesis and consequently root and shoot growth. Many soybean cultivars can accumulate high Cd concentration in seed when grown on Cd-polluted soil [19, 20]. Consumption of food containing excessive Cd leads to a risk of chronic toxicity. In humans, it can damage kidneys, causing a loss of calcium and associated osteoporosis [21]. To reduce the health risk, it is desirable to limit the concentration of Cd in crops used for human consumption. Due to growing concern about safety of foods and human health, the Codex Alimentarius Commission of Food and Agriculture Organization/World Health Organization (FAO/WHO) has proposed an upper limit of 0.2 mg kg^{-1} for Cd concentration in soybean grain [22]. However, a large-scale survey of agricultural products revealed that the Cd concentration of 16.7% of soybean seeds exceeded the international allowable limit of 0.2 mg kg^{-1}, which is much higher than that of other upland crops [23]. Cultivars with reduced uptake of Cd are needed for human consumption. Cd uptake depends both on the Cd concentration in the soil and on the characteristics of the specific cultivars. Breeding cultivar with reduced Cd is an attractive method for changing the element profile of crops as the benefit will persist in the seed that can reduce the requirement for other management practices [24].

4. Nutritional factors

4.1 Soybean protein

Soybean contains about 40% protein and is noteworthy as it is the most complete vegetable protein [25]. Concretely, with exception of sulfur-containing amino acids such as methionine, the amino acid pattern of soybean resembles the pattern derived from high-quality animal protein sources [25]. In fact, soybean protein can even enhance the nutritional quality of other vegetable protein. Protein sources that are deficient in some amino acids can be complemented by soybean. Soybean is rich in lysine, tryptophan, threonine, isoleucine, and valine and therefore complements well with cereal grains that are deficient in those amino acids [26]. By ultracentrifugation studies, four different fractions have been revealed, with approximate Svedberg coefficients of 2S, 7S, 11S, and 15S [6]. The 2S fraction contains from 8 to 22% of the extractable soybean protein. It consists of several enzymes, including the trypsin inhibitors, Bowman-Birk and Kunitz inhibitors [6]. Trypsin inhibitors inhibit the protein-cleavage effect of proteases (such as trypsin) affecting the digestibility and leading to growth depression in animals. Therefore, soybean meal needs first to be heated in order to inactivate the trypsin inhibitors. However, trypsin inhibitors have been found to be powerful anti-carcinogenic agents in humans and therefore they can be considered as functional components of soybeans [27].

More than 70% of the soybean seed storage protein is composed of 7S β-conglycinin and 11S glycinin. The 7S fraction makes up 35% of the extractable soybean protein. The quantity and the quality of the protein in the seed are the major biochemical components influencing the quality of tofu and other soy food products [28]. The mean glycinin to β-conglycinin protein ratio is known to influence the protein quality of soybeans, and greatly affects the functional properties of food products made from soybeans [29, 30]. Glycinin and β-conglycinin also differ in amino acid composition, with glycinin being higher in sulfur (S), containing amino acids that account for 3–4.5% of the total amino acid residues [31]. G1, G2, and G7 glycinin subunits contain a higher amount of methionine (6–7 per subunit) compared to G3, G4, and G5 glycinin subunits, which contain 5, 2, and 4 methionine residues per subunit, respectively [31]. By comparison, β-conglycinin is devoid of methionine [32, 33] and β-conglycinin contains a major allergen in its subunit [34]. Increased glycinin content in soybean protein is an important trait for increasing the concentration of the S-containing amino acids [35]. Because glycinin and β-conglycinin have a great impact on the nutritional value and quality of soybean products, these two storage proteins have been extensively studied and targeted for genetic manipulation in breeding programs. Soybean mutant genotypes differing in seed storage glycinin and β-conglycinin subunit composition were developed and tested for their effects on tofu quality [30]. It was shown that group IIb (A_3) glycinin played the major role in contributing to tofu firmness with any coagulant, while the group IIa (A_4) subunit could have a negative effect on tofu quality. Yu et al. [36] reported that soybean cultivars with 7S α′ and 11 S a4 nulls always make firm tofu than the check cultivar Harovinton. The hardness of gels from glycinin decreased in the order of group IIa, IIb, and I [37, 38]. Protein subunit composition also affects the quality and stability of soymilk [39].

Other soybean seed proteins include lipoxygenase and lectins. The lipoxygenase enzyme constitutes about 1–2% of the soybean protein. The lipoxygenase enzyme generates a grassy-beany flavor when it oxidizes fats and is not preferred by consumers in some countries. It is possible to avoid the oxidation of the fats by heat inactivation of the lipoxygenase enzyme; however, this is cost-ineffective and leads to insolubilization of proteins. Therefore, the genetic elimination of the lipoxygenase is preferred in order to reduce the beany flavor. Genotypic variation and the influence of growing environment on lipoxygenase accumulation in soybean seed are well documented in the literature [2, 26]. Lipoxygenase 1, 2, and 3 null germplasm lines were developed and showed that the grassy-beany flavor was eliminated [40]. Triple-null soybeans can be used for edible soy products, such as soymilk and tofu [40]. Similarly, saponins and isoflavones may also be the cause of undesirable taste in soy products although this is not well documented yet. The breeding of cultivars with low isoflavones and saponins is possible [2]. The 11S fraction comprises 31–52% of the extractable soybean proteins [6]. The 11S fraction is responsible for the gelling character of tofu, and hence, the proportion of this fraction compared to 7S plays an important role in tofu firmness [2]. The 15 S fraction comprises about 5% of the total extractable protein. It is only poorly characterized and is thought to be composed of polymers of the other soybean proteins [6].

4.2 Carbohydrates

Dry soybeans contain on average 35% of carbohydrates, which can be divided into soluble and insoluble carbohydrates [27]. Soybean seeds possess 15–20 different soluble carbohydrates that makes up approximately 15–25% of dry weight [41]. Sucrose, raffinose, and stachyose are the most relevant soluble carbohydrates for breeding of food-grade soybean. Sucrose in dry soybean seeds is found in contents of typically 5.5% [27]. Sucrose is important for improving taste in soybean-based products. The oligosaccharides raffinose and stachyose typically constitute about 0.9 and 3.5% of dry soybean seeds, respectively [27]. The seed coat of soybeans contains a major part of insoluble carbohydrates such as cellulose, hemicellulose, pectin, and a trace amount of starch [27]. Consumers, especially in countries where fermented and vegetable soybean are not in vogue, may be skeptical toward the use of soy products because of flatulence and poor digestibility. These effects are caused by oligosaccharides, stachyose and raffinose. Humans and monogastric animals do not possess the enzyme called α-galactosidase necessary for hydrolyzing the linkages present in these oligosaccharides, so they cannot be digested when consumed. Intact oligosaccharides reach the lower intestine and undergo anaerobic fermentation by bacteria with gas expulsion (H_2, CO_2, and traces of CH_4), causing the flatus effect and sometimes diarrhea and abdominal pain. Although raffinose and stachyose can be reduced to an extent by soaking or boiling, genetic reduction is one of the prime plant breeding objectives.

4.3 Soybean oil

The major components of crude soybean oil are triglycerides. After refinement of the oil, soybean oil is composed of 99% of triglycerides. Triglycerides are neutral lipids composed of one glycerol linking three fatty acids [27]. The saturated fatty acids in soybean oil are palmitic acid (16:0) and stearic acid (18:0), with average concentrations of about 11 and 4% (relative to the oil), respectively, and they are useful in making low trans-fat margarines. Soybean oil contains an average of

22% monounsaturated fatty acid, oleic acid (18:1). Monounsaturated fatty acids are healthy and have good oil stability [42]. Soybean oil possesses the two polyunsaturated fatty acids: linoleic acid (18:2), an omega-6 fatty acid, and linolenic acid (18:3), an omega-3 fatty acid [26]. They can be found in average concentrations of 53 and 8% of the oil, for linoleic and linolenic acid, respectively. Low (reduced) linolenic soybeans have half the linolenic acid level of standard soybeans, which reduces the need for hydrogenation, a process used in converting vegetable oils to margarine that results in the production of unhealthy trans fatty acids.

Soybean crude oil is also shown to consist of phospholipids, unsaponifiable material, free fatty acids, and metals. Unsaponifiable material consists of tocopherols, phytosterols, and hydrocarbons [27]. Tocopherols and phytosterols are considered as functional components. Soybean oil provides an additional benefit due to presence of enriched amounts of α-tocopherol or natural vitamin E. Oils containing low contents of linolenic acid (18:3) have been shown to contain high amount of α-tocopherol and results in lowered amount of ϒ-tocopherol [5].

4.4 Vitamins and minerals

Soybeans contain water-soluble and oil-soluble vitamins. The water-soluble vitamins such as vitamin B1 (thiamin), vitamin B2 (riboflavin), vitamin B5 (pantothenic acid), and vitamin B6 (niacin) and the oil-soluble vitamins vitamin A and vitamin E (tocopherols) are present in soybean. Vitamin A mainly exists in the form of β-carotene in immature and germinated seeds, whereas it is present in negligible amount in mature seeds [27]. Most of the minerals are found in the meal fraction rather than in the soybean oil fraction. Dry soybean seeds contain on an average concentration ranging from 0.2 to 2.1% major minerals such as potassium, which is present in the highest concentration followed by phosphorus, magnesium, sulfur, calcium, chloride, and sodium [27]. Minor minerals found in soybeans include silicon, iron, zinc, manganese, copper, molybdenum, fluorine, chromium, selenium, cobalt, cadmium, lead, arsenic, mercury, and iodine [27].

4.5 Functional components

Functional components of soybeans include isoflavones, saponins, lecithin, trypsin inhibitors, lectins, oligosaccharides, tocopherols, and phytosterols [27]. Presence of such biological ingredient creates interest to consider soybean food products as functional foods, i.e., foods that contain biological components that deliver special health benefits, e.g., anticancer, hypocholesteromic, and antioxidative effects to the consumer [26]. Isoflavones are phytoestrogens and are known to have positive health effects such as the reduction of the risks for coronary heart disease, osteoporosis, certain types of cancer, and the moderation of postmenopausal symptoms in women [43]. Soybean possesses 0.1–0.4% of isoflavones on a dry weight basis; hence, soybean possesses the highest amount of isoflavones compared to all other crops [27]. The isoflavone concentration varies considerably depending upon the genotype and environmental conditions. It is thought that isoflavones are mainly responsible for most of the health benefits from soybean-based foods. Therefore, they gained more and more attention from the scientific world [27], and research on breeding for enhanced isoflavone content is increasing. Refined soybean oil possesses about 1000–2000 mg/kg. Tocopherol exists in four isomers, three of them being α-, ϒ-, and δ-isomers that are present in soybean oil. α-tocopherol (natural vitamin E) in soybean is the leading commercial source of this vitamin. Tocopherols protect the polyunsaturated fatty acids from oxidation; hence, they are antioxidants and used in pharmaceutical applications [42].

5. Soyfoods

5.1 Tofu

Soybeans with large seed size and high protein levels are primarily used for soymilk and tofu production. Other traditional food from soybean includes tempeh, miso, soy sauce, okara, soynuts, soy milk, yoghurt, meat, and cheese alternatives. Tofu is perhaps the most widely consumed soy food in the world. Tofu is naturally processed and it retains a good amount of nutrients and phytochemicals such as the isoflavones [5]. Tofu typically contains 7.8% protein and 4.2% lipid on a wet basis [5]. It has a relatively low carbohydrate and fiber content, making it easier to digest. There are two main types of tofu: silken, or soft tofu and hard tofu. They are made by soaking whole soybeans and grinding them into a slurry with water. The slurry is cooked to form soymilk and a coagulant is added. The most commonly used coagulants are magnesium chloride, calcium sulfate, or glucono-D-lactone; the coagulants can be used purely or in combinations to achieve different flavor or textural characteristics. Heating is also usually applied in order to facilitate the coagulation. The result of the coagulation is that after a few minutes, the soymilk begins to curdle and large white clouds of tofu curd are formed. The water in the curds are then removed and placing the tofu curd in cloth-lined forming boxes where pressure is applied from the top results in the formation of hard tofu. Silken tofu in comparison to hard tofu is not pressed and is often coagulated in the container in which it is to be sold. [2].

5.2 Soymilk

The popularity of soymilk has expanded from Asia to the U.S. and Europe since the 1980s. Traditionally, it is made from whole beans in the same way as the first few steps of tofu manufacture. This soy milk contains nutrients, saponins, isoflavones, and other soluble components of the soybean from which the soy milk is made. Some manufacturers add isoflavones back into the soy milk in order to make health claims about the product. Additionally, soymilks are also fortified with vitamins and minerals, such as β-carotene and calcium or docosahexaenoic acid (DHA), an omega-3 fatty acid [2]. However, beverage-quality soy milks available in the market are usually prepared from soy protein isolate, to which sugars, fats, and carbohydrates are added to improve flavor and generate a nutritional profile similar to that of cow's milk [2].

5.3 Vegetable soybeans (edamame, mukimame)

Vegetable soybean consists of the whole soybean picked at the R6-R7 stage and seeds are bigger and sweeter. At this stage, the soybean has a firm texture, contains a high level of sucrose, chlorophyll, and is at its peak of green maturity. The harvested pod can be left entire or be shucked into individual beans. After being blanched and frozen, the soybean can be sold as "edamame," referring to the entire pod, or "mukimame," referring to individual beans [2]. Nutritionally, it is highly rich in protein (11–16%), monounsaturated fatty acid, vitamin C, fiber, iron, zinc, calcium, phosphorous, folate, magnesium, potassium, tocopherol, and anticancer isoflavones [44]. It also has a pleasant flavor and soft texture and is easier to cook. Cooked vegetable soybean has the highest net protein utilization value (NPU: ratio of amino acid converted to protein) among all soy products. Vegetable soybean also has 60% more calcium and twice the phosphorus and potassium levels of green peas, which is India's most commonly consumed fresh legume (https://www.gov.uk/government/case-studies/dfid). The vegetable soybean, in general, carries a flavor, called "beany flavor" or "grassy flavor." Genotypes with high levels of sucrose, aspartic acid, glutamic acid, and alanine are found to have acceptable taste [44]. Biochemical analysis has established that production of "beany flavor" in soybean or soy-based products is primarily due to the lipoxygenase or the oxidative rancidity of unsaturated fatty acids [45]. Plant lipids are sequentially degraded into volatile and nonvolatile compounds by a series of enzymes via the lipoxygenase pathway, which catalyzes the hydroperoxidation of polyunsaturated fatty acids to form the aldehyde and alcohols that are responsible for the grassy-beany flavor [46]. *Organic food-grade* soybeans are produced using cultivation practices that do not use synthetic compounds. In the U.S., growers producing and selling soybeans that are labeled "organic" must be certified by a USDA-approved state or private agency. The top selling organically produced soy products in the US are tofu and soymilk. Other specialty soybeans include varieties with low saturated fat, high isoflavone, high sucrose, high oleic acid, high stearate, or high protein. Large-seeded soybeans with thin seed coat and a clear hilum are preferred for the soynut market, while small- to medium-sized seeds are preferred for sprouts.

5.4 Breeding targets for food grade soybeans

Breeding for food-grade soybeans with unique seed composition has focused on a specific nutritional trait of the soybean seed. Examples of such varieties are given according to the fraction from which the targeted trait origins. Food-grade soybean that targets a specific trait such as varieties high in total protein content, high in β-conglycinin, low in lipoxygenase, high in specific amino acids such as lysine, methionine, and threonine, and low in allergenic proteins [13]. High-protein soybeans (>43%) are used for tofu, soymilk, soy sauce, beverages, baked goods, pudding, cheese, and meat analogs. The breeding of food-grade soybeans can be classified into three major categories: the breeding of large-seeded soybeans, the breeding of small-seeded soybeans, and the breeding of soybean with unique seed composition [13].

5.5 Large-seeded soybeans

By targeting specific traits, soybean breeders try to develop soybeans with good yield and quality [5]. Large-seeded soybeans are bred for tofu, soymilk, miso, edamame, and soynuts [13]. An important factor for the breeding of tofu soybeans is the tofu yield, which is defined as the weight of fresh tofu produced from a unit of harvested soybean. Seed size and seed appearance are also of importance for tofu soybeans. Tofu soybeans are larger than 20 g/100-seeds [13]. It is possible to produce good quality tofu with dark hilum beans but this requires prior dehulling of the beans and careful soymilk filtration [5]. In order to avoid these additional processing steps, soybeans with a yellow cotyledon, yellow seed coat, and clear hilum are preferred. Moreover, a thin but strong seed coat that is free from cracking and discoloration is desirable [13]. Soybean seeds with high protein content exceeding 45% on dry matter basis and improved ratio of 11S/7S is desirable for tofu soybeans as this enhances tofu yield and gelling characteristics, respectively [5]. A high protein/oil ratio provides a higher tofu yield and firmer texture; therefore, low oil content is preferred. Moreover, tofu soybeans should have high water uptake, a low calcium content, and a high germination rate. The carbohydrate content and composition influence the taste of tofu and soymilk [13]. High total sugar content (above 8% on dry matter basis) [5], high sucrose, low raffinose, and low stachyose are highly desirable for tofu and soymilk [13]. Examples of tofu and soymilk varieties: Black Kato, Toyopro, Grande, Proto (from Minnesota), Vinton-81, HP 204, IA1007, IA1008 (from Iowa), and Harovinton [13, 47].

5.5.1 Vegetable soybean

Vegetable soybean varieties should meet certain requirements such as sweeter seeds with thin seed coat and large seed size (>30 g/100-seeds dry weight) [13]. As the pods are eaten directly, genotypes with sparse gray pubescence with green and thin seed coat are preferred [13]. Moreover, edamame cultivars should possess as less as possible of one-seeded pods as they require greater effort to shell by consumers. Those cultivars with genetically "stay green" and delayed yellowing toward maturity make it possible for growers to have extended harvest period closer to maturity. Vegetable-type soybean should possess important nutritional traits such as high content of sugar (sucrose and maltose) and free amino acids to impart sweet and delicious taste. Sucrose is primarily responsible for the sweetness of vegetable soybeans, where sucrose content is preferably higher than 10% on dry matter basis. Certain free amino acids, such as glutamic acids, are major contributors to the taste of vegetable soybeans [13].

5.6 Small-seeded soybeans

5.6.1 Natto

Natto beans are small-seeded soybeans typically used for the fermented soy foods popular in Japan. For natto, small to ultrasmall soybeans (smaller than 9 g/100-seeds [2] of maximum of 5.5 mm diameter are preferred for better fermentation. The seeds have preferably a near spherical shape as this reduces the ratio of the tough seed coat to softer cotyledon [2]. Also, clear hilum and thin seed coat are desirable traits for natto soybeans. Natto soybeans are nutritionally characterized by a high carbohydrate content [13]. A high content of soluble sugars (>10%) on a dry weight basis results in a softer natto product, an important requirement for natto [5]. The composition of sugars is important for the effectiveness of fermentation [13]. To obtain a steady and controlled fermentation, low sucrose content with high stachyose and raffinose content is favored [5]. Soybean with moderately high protein content is desirable in order to provide amino acids for fermentation. Oil content must be low, i.e., less than 18% of the dry matter as it enhances water absorption [13]. For a softer natto product, seeds must additionally possess high water absorption capacity during soaking, which is the first step of natto manufacturing. Breeders use standard small-seeded lines, such as the cultivar Vance (known for having a medium ability for water uptake), to compare selected lines for water absorption capacity [5]. Soybeans with medium seed size (10–12 g/100-seeds) and a high germination rate are preferred for bean sprouts. High-protein, high-isoflavone, high-sugar, and lipoxygenase-free soybeans are desirable for soybean sprouts [13].

It is reported that the Asian small-seeded lines exhibited high diversity indices than the U.S. lines for seed hardness, calcium content, and stone seed rate. In addition, the average genetic diversity of the U.S. small-seeded soybeans (1.48) was lower than that of Asian small-seeded soybeans (1.57), suggesting narrower genetic base in the U.S. lines. Seed uniformity, hardness, protein, and calcium content appeared to be relatively high in diversity index for both the U.S. and the Asian large-seeded lines. The U.S. small-seeded soybeans were desirable for natto production because of their softer texture with higher water absorption capacity and lower stone seed ratio. However, the Asian large-seeded soybeans had a lower stone seed ratio and a higher water absorption capacity. Therefore, using the Asian large-seeded genotypes may potentially improve seed quality for tofu and soymilk [48]. Therefore, the Asian soybean gene pool may serve as valuable genetic source for increasing protein content of the U.S. food-grade soybeans.

6. Breeding for protein content

Availability of genetic variability for soybean food-grade traits offers scope to improve through breeding. Breeding cultivated soybean varieties with high protein or high oil are an extremely important and promising objective. High protein and low oil content add nutritional value to soy foods. Germplasms that cover a wide range in protein content (33.1–55.9%) and oil content (13.6–23.6%) are available for breeders to modify the seed/oil ratio in the breeding program. The negative correlation between protein and oil facilitates the development of high protein and low oil lines. High protein content is generally associated with low yield, which makes the development of lines that combine high protein and high yield difficult. However, high yield is mostly achieved by selection for moderately high protein content (43–45%) [13]. Seed protein and oil content are two valuable quality traits controlled by multiple genes in soybean. The phenotypic range of protein content of soybean has been reported to be 34.1–56.8% of seed dry mass, and oil content ranged from 8.3 to 27.9% [49], suggesting that there is great potential for genetic improvement of soybean seed protein and oil content. The negative correlation between oil and protein content makes improvement of both traits simultaneously a challenging task using conventional breeding [50]. Therefore, the identification of molecular markers associated with quantitative trait loci (QTLs) controlling protein and oil content is a prerequisite for breaking the negative correlations between both traits [51].

In the SoyBase database, 241 QTLs for protein content and 315 QTLs for oil content were reported and found to be distributed over 20 soybean chromosomes [52]. A majority of these QTLs were mapped by linkage mapping based on bi-parental populations and limited by the relatively small phenotypic variation and by the fact that only two alleles per locus can be studied simultaneously. The broad chromosome regions of QTLs make it difficult to identify putative candidate genes of interest [53]. With the advancement of genetic map construction, the availability of a well-annotated reference genome, resources for association mapping, and whole-genome resequencing (WGRS) data, a large number of QTLs for seed protein content have been identified (**Table 1**).

Several genome-wide association studies [50, 66, 68] and QTL analysis [53, 56] have shown similar QTL genomic loci (e.g., Chrs20, 15, and 5) for protein and oil indicating negative pleiotropic effect or linkage (larger LD). The QTL on Chr20 was

Trait	Major QTL (Chr.)	References
Seed protein	20, 15, 18	Akond et al. [54]; Diers et al. [55]; Nichols et al. [56]; Brummer et al. [57]; Lee et al. [58]; Sebolt et al. [59]; Chung et al. [60]; Fasoula et al. [61]; Panthee et al. [62]; Lu et al. [63]; Pathan et al. [53]; Wang et al. [64]; Phansak et al. [65]; Hwang et al. [50]; Vaughn et al. [66]; Sonah et al. [67]; Bandillo et al. [68]; Valliyodan et al. [69]; Zhou et al. [70]; Tajuddin et al. [71]
	5, 6, 8	Hyten et al. [72]; Pathan et al. [53], Sonah et al. [67]; Bandillo et al. [68]
	10, 14	Wang et al. [64]; Phansak et al. [65]
	13, 03, 17, 12, 11	Zhou et al. [70]
Sucrose and oligosaccharides content	5, 7, 8, 11, 13, 18, 16, 19, 20	Maughan et al. [73]; Kim et al. [74]; Kim et al. [75]; Alkond et al. [76], Cicek [77]; Clevinger [78]; Maroof et al. [79]; Feng et al. [80]; Zeng et al. [81]; Wang et al. [82]

Table 1.
Major QTLs for seed protein, sucrose and oligosaccharide content reported in soybean.

most likely in the genomic region of 29.8–31.6 Mbp that was supported by integrating GWAS, transcriptome, and QTL mapping analysis (**Table 1**) [68]. It was observed that the gene order was conserved and 18 identified genes were tandemly duplicated on Chr10 and showed similar gene ontology [83]. Three putative candidate genes were identified on Chr20 and suggested that these non-duplicated genes might be related to protein content [68]. Similarly, Chr15 QTL (38.1–39.7 Mbp) showed an inversely duplicated genomic block on Chr8. The QTL on Chr15 comprises 18 putative genes, 13 of which were duplicated with similar gene function. Syntenic analysis provided a basis for divergence of QTL regions that took place during recent genome duplication and suggested the retention or loss of several genes that might be responsible for oil content and protein in soybean. In addition to pleiotropic effects of protein on oil and yield, variation in seed protein concentration significantly affects seed size, crop growth, and development [84]. High-protein genotypes showed lower leaf area and harvest index when compared with high-yielding genotypes. While high-protein small seed showed higher leaf area at the beginning of seed fill, more canopy biomass production, and low levels of assimilate per seed [84]. Therefore, breaking the undesirable genetic linkage between protein, oil, and yield related loci through repetitive recombination and random mating is necessary.

7. Breeding for 11S/7S ratio

Consumers have preference for firmer tofu texture that partly depends upon the protein composition. The genotypic variation in this trait is partly due to the ratio of 11S-to-7S protein fraction in the seed. The 11S fraction generally possesses greater gelling potential than 7S; hence, high 11S-to-7S ratio is desirable as it results into harder than those with low ratio. The 11S-to-7S ratio is reported to range from 0.3 to 4.9. However, genotypes with same 11S-to-7S ratio do not always result in the same firmness because of different 11S subunit composition. In general, a high 11S-to-7S ratio as well as suitable 11S composition is of importance for good tofu firmness.

The selection and manipulation of specific subunit composition will play a major role in the development of improved protein quality. Molecular markers linked to the various subunit of glycinin and β-conglycinin have been reported previously. PCR-based markers were reported for the identification of β-conglycinin genes [85, 86]. An RFLP marker associated with the *Scg-1* (suppressor of β-conglycinin) gene was developed by using the α-subunit gene as probe [87]. SNPs in the β subunit genes were used to map the *Scg-1* gene, and the chromosomal region associated with β-conglycinin deficiency was located on linkage group I of the soybean genetic map [86]. Hayashi et al. [88] reported AFLP markers linked to the recessive allele, cgdef, controlling the mutant line lacking 7S globulin subunits (α, α′, β). Markers linked to the glycinin genes were reported. RFLP markers were identified for both *Gy*4 and *Gy*5 and mapped in linkage group O and F on the public soybean linkage map [31, 89]. *Gy*1, *Gy*2, and *gy*6 are linked in tandem to one another on linkage group N, while *Gy*3 and *Gy*7 are linked to one another on linkage group L [90]. KASP-SNP markers linked to 7S α′ and 11S A_1, A_3, and A_4 subunits have been reported [91]. Three SSR markers (Satt461, Satt292, and Satt156) were found to be associated with glycinin QTLs that were distributed on linkage group D_2, I, and L, whereas two β-conglycinin QTL-associated SSRs (Satt461 and Satt249) were distributed on LG D_2 and J [35].

Functional markers (FMs) have advantages over the linked markers, because their polymorphic sites have been derived from the genes involved in phenotypic trait variation [92]. Glycinin genes have high degree of conservation within the subgenus *Soja*, but there are more variations within subgenus *Glycine* [93]. Despite the high degree of similarity among the subunits in Group I and Group II, gene primer pairs specific to *Gy1*, *Gy2*, and *Gy5* were designed [93–95]. PCR primers were designed for identification of the *Gy4* null allele and demonstrated selection of soybean without the A_4 peptide can be done by null allele specific primers which overcome the drawbacks of SDS-PAGE gel-based selection for A_4 peptide [95].

8. Breeding for amino acid composition

Besides breeding for increased protein content, protein composition is important for its nutritional value. Based on solubility properties, globulins and albumins are two major components of dicot seed storage protein, and soybean primarily belongs to the globulin (~70%) family [96]. The soybean globulins (glycinin and β-conglycinin) are relatively low in sulfur-containing amino acids methionine (Met) and cysteine (Cys) as well as threonine (Thr) and lysine (Lys) [97]. Increasing the soybean storage protein content of seed along with improving the ratio of glycinin to β-conglycinin is of great potential for food grade soybean improvement [98, 99]. Therefore, besides increased protein content, enhancing sulfur containing amino acids (Met, Thr, Cys, and Lys) would improve the nutritional value. More than 70% of the essential amino acid enriched meal is used in the feed industry [97, 100]. Although soybean cultivars with improved protein content have been successfully developed, only a few studies have been conducted to identify genomic regions controlling amino acid composition. The difficulty in breeding for improved amino acids could be due to lack of genetic variability, lack of high throughput, and cost-effective phenotyping platform to screen a large number of samples for amino acids. Panthee et al. [99] identified QTL for essential amino acids in a F_6-derived recombinant inbred population. In another study, a major QTL for essential amino acids and crude protein was identified on Chr20 [97]. Moreover, negative correlations of crude protein with Lys and Thr and a positive correlation between Thr with Lys were also observed [97]. Among the essential amino acids, Met, Lys, and Thr are synthesized from a common precursor aspartate; thus, they are strongly correlated. Krishnan et al. [101] introgressed leginsulin (Cys-rich protein) and a high protein trait from an Asian soybean germplasm, PI 427138, into North American experimental line (LD00–3309). While they were successful in introgressing leginsulin and improving protein content, the overall concentration of sulfur-containing amino acids was not changed compared to parental lines.

Seed protein content and composition are dependent on the genetic background of an elite parent that plays an important role in the expression of a newly introgressed allele because of complex epistatic interactions [102]. It has been found that most of the QTLs affecting seed protein and yield and yield-related components were detectable only in one of the parental genetic backgrounds (GBs) in introgression lines of reciprocal crosses [103]. The high protein allele within a different genetic background resulted into reduced Thr and Lys content [103]. The high protein allele from Danbaekkong on Chr20 has been demonstrated to increase seed protein content in several maturity groups (III–VIII) in various genetic backgrounds with little drag on seed yield [104]. On the other hand, yield drag was observed for the protein QTL alleles on Chr20 from other sources, including wild *G. soja* [56, 60]. Hence, it is not feasible to select only the major crude protein QTL on Chr20 to improve protein quality. Improvement of protein and amino acid profiles has been limited by the narrow genetic base and genome complexity of soybean. Mutation breeding can be used to enhance the genetic variability. Mutagenized populations (physical, chemical, transposon tagging or transformation-induced mutagens) have been useful in crop improvement [105]. In soybean, mutations for seed traits, including oleic acid [106], oil [105, 107], stearic acid [108], and lipoxygenase [109] were identified using induced mutation.

9. Genomics-assisted breeding (GAB)

The integration of genomic tools and breeding practices are the core components of genomics-assisted breeding (GAB) for developing improved cultivars for any given trait. Near-isogenic lines (NILs) can be developed for major QTL (e.g., protein QTL on Chr20) by backcross breeding. Using NILs, the effect of a QTL and the phenotype it produces (i.e., protein or amino acid content) can be estimated precisely without the confounding effects of differences in genetic backgrounds. Additionally, developing NILs in a range of maturity groups is desirable to study the effect of environment and maturity on seed protein content. Marker-assisted backcrossing selection approach was utilized to produce a NIL-(cgy-2–NIL)-containing mutant cgy-2 allele, responsible for the absence of allergenic α-subunit of β-conglycinin [110]. It is also possible to incorporate multiple genes/QTL into elite lines in a cyclic forward crossing scheme and employing marker-assisted recurrent selection (MARS) as an effective approach [111, 112]. Recurrent selection was effectively utilized for increased gain yield, protein, oil, and oleic acid content [111, 113, 114]. Furthermore, the next-generation sequencing (NGS) data can be used effectively for genomic selection (GS) to identify desirable parents and progenies. Jarquin et al. [115] assessed the genomic and phenotypic data of over 9000 accessions and developed genomic predication models to evaluate the genetic value for protein, oil, and yield traits. Similarly, genomics-assisted haplotype analysis is a promising approach if the information of a major QTL is available and that can be applied to select desirable haplotype blocks for parental selection and crossing by design [116].

In order to widen the genetic base, it may be necessary to utilize wild species accessions as introgression libraries as well as developing interspecific populations. On the other hand, elite cultivars and landraces can be used to develop mapping populations, and training populations [114]. Wild soybean (*G. soja*) serves as a unique resource to study regulation of protein and amino acid biosynthesis, because the seed concentration of these components is higher in *G. soja* compared with *G. max.* Utilization of *G. soja* in breeding program is hampered due to linkage drag on favorable agronomic characteristics [113]. However, this issue could be resolved by advanced backcross QTL-based breeding, which was utilized for introgressing alleles from wild tomato to cultivated type for yield improvement [117], or through mutation breeding approaches.

10. Breeding for carbohydrate content in soybean seeds

10.1 Sucrose content

Breeders aim to increase the sucrose content in soybean seeds which contribute to the sweet taste of soy foods, especially for tofu, soy milk, and edamame. The sucrose content in soybeans ranges from 1.5 to 10.2%, and germplasm with even higher content, 13.6%, has been identified [13]. Varieties that target a specific component of the carbohydrate fraction are varieties high in sucrose content and varieties low in oligosaccharides [13]. Compared to conventional soybeans, high-sucrose soybeans contain 40% more sucrose but 90% less stachyose and raffinose. High-sucrose soybeans are used to produce tofu, soymilk, beverages, baked goods, puddings, cheese, and meat analogs [13]. The genotypic correlation between sucrose and 100-seed weight is positive and significant, as well as the genotypic correlation of 1000-seed weight with protein. Moreover, the heritability for 1000-seed weight is high. Hence, the breeding program selection on 100-seed weight would result in a good response on relative protein and sucrose content.

10.2 Oligosaccharides content

Stachyose and raffinose are not readily digestible and cause flatulence when soy foods are consumed. Therefore, breeders aim to develop soybean seeds with reduced oligosaccharide content. Stachyose and raffinose content among soybean germplasm range from 1.4 to 6.7%, and 0.1 to 2.1%, respectively. Breeding lines with less than 1% stachyose and raffinose have been developed [13]. Soybean germplasm "V99–5089" was developed with high sucrose, low raffinose, and low stachyose content to use as a parent in food-grade soybean breeding programs [118]. The genetic variability of seed sugars has significant allelic difference in the genes controlling the biosynthetic enzymes. QTL mapping of soluble sugars in soybean seed were reported and of which 28 were for seed sucrose (**Table 1**). These 28 QTLs were mapped on LGs A1 and E; 3 QTLs on A2, I, and F, and 3 QTLs on L, M, and B1 [73], two QTLs on L, D1b, 7 QTLs on L [74], and B2, D1B, E, H, J [75]. The genomic regions associated with sucrose, raffinose, and stachyose were identified in segregating $F_{2\text{–}10}$ RILs [74].

Alkond et al. [76] reported 14 significant QTLs associated with sucrose and oligosaccharides that were mapped on 8 different linkage groups (LGs) and chromosomes (Chr). Seven QTL were identified for raffinose content on LGs D1a (Chr1), N (Chr3), C2 (Chr6), K (Chr9), B2 (Chr14), and J (Chr16). Four QTL for stachyose content were identified on LG D1a (Chr1), C2 (Chr6), H (Chr12), and B2 (Chr14) [76]. Three QTL for seed sucrose content were identified on LGs N (Chr3), K (Chr9), and E (Chr15). The region of Chr15 (LG E) that has been reported to be associated with sucrose was detected by others [73, 75, 77, 78], but the position of the QTL was different [76]. The two of the regions underlying seed sucrose QTLs identified on LG N (Chr3) and K (Chr9) are additions to the loci previously reported on LGs D1b (Chr2), A1 (Chr5), M (Chr7), A2 (Chr8), B1 (Chr11), H (Chr12), F (Chr13), G (Chr18), J (Chr16), L (Chr19), and I (Chr20) [73, 75, 77, 78]. The selection for beneficial alleles of these QTLs could facilitate breeding strategies to develop soybean lines with higher concentrations of sucrose and lower levels of raffinose and stachyose.

11. Breeding for functional components

11.1 Lipoxygenase

Normal soybean seeds contain three lipoxygenase isozymes that are responsible for the grassy beany flavor and bitter taste of soy food. Research is being conducted for the genetic elimination of lipoxygenase from soybean seeds to reduce undesirable flavors in soy food products. Soybean seed lipoxygenase exists in three isozymic forms, namely lipoxygenase-1, −2, and −3 controlled by single dominant genes, viz. *Lx1*, *Lx2*, and *Lx3*, respectively. Their recessive forms, i.e., *lx1*, *lx2*, and *lx3* cause the loss in activity of corresponding isozyme [119]. Several combinations of lipoxygenase null mutants have already been developed: 0-, 00-, and 000-genotypes with one, two, and three of the isozymes eliminated respectively, In the 000-genotype, absence of the grassy and beany flavor was observed, as there was no detectable level of the lipoxygenase proteins in mature soybean seeds. The presence or absence of three lipoxygenase isozymes is determined by gel electrophoresis and spectrophotometer or by immunological or colorimetric methods [13]. Of the three lipoxygenases, *Lx2* locus has been mapped on *chr13*, which corresponds to linkage group F, and has been reported tightly linked with *Lx1* locus [120]. *Lx3* gene has been reported to be present on *chr15* and is inherited independent of *Lx1* and *Lx2*. SSR marker Satt656 tightly linked with Lox2 [121] has been deployed in the development of Lox-2 free soybean genotypes NRC109 and NRC110 in India [121].

Based on *Lx3* mutant gene sequence, SNP (Lox3PM1) and STS marker (Lox3–3′) were developed for identification of Lox*3* null individual [122].

11.2 Isoflavones

Soybean cultivars with good isoflavone content are desirable as it contributes health benefits. High-isoflavone soybeans contain more than 0.4% isoflavones compared to levels of 0.15–0.25% for traditional soybean varieties [13]. Isoflavone content is influenced by genetic factors and environmental factors such as temperature and irrigation during seed maturation [13]. For instance, the total isoflavone content of soybean seeds appears to be negatively related to growth temperature [5]. Understanding the genetic regulation of this pathway may be necessary for obtaining cultivars with good isoflavone levels. Interest has been put in the phenylpropanoid synthetic pathway which is catalyzed in its first step by isoflavone synthase (IFS). Two genes for IFS have been identified in soybean. Furthermore, negative correlation has been found between total isoflavone content and linolenic acid (18:3) concentration. Other data suggest negative correlation between isoflavone content and protein content [5]. QTLs affecting isoflavones were identified using recombinant inbred line population and found five QTLs contributed to the concentration of isoflavones, having single or multiple additive effects on isoflavone component traits [123]. Similarly, six QTLs were identified using the linkage map constructed with specific length amplified fragment sequencing, of which one major QTL (qIF20-2) contributed to a majority of isoflavone components across various environments and explained a high amount of phenotypic variance (8.7–35.3%) [124]. Akond et al. [125] identified QTL controlling isoflavone content in a set of recombinant inbred line (RIL) populations of soybean derived from "MD96–5722" by "Spencer" cultivars. Wide variations were found for seed concentrations of daidzein, glycitein, genistein, and total isoflavones among RIL populations. Three QTLs were identified on three different linkage groups (LG). One QTL that controlled daidzein content was identified on LG A1 (Chr 5) and two QTLs that underlay glycitein content were identified on LG K (Chr 9) and LG B2 (Chr 14). Identified QTLs could be used to develop soybean with preferable isoflavone concentrations in the seeds through MAS.

11.3 Seed oil concentration

Increasing the seed oil concentration has been a breeding goal for centuries. The ancestor of the domesticated soybean used to have small, hard, black seeds with low oil content, high protein content, and low yield. It is known that an increase in oil content is positively correlated with yield and negatively correlated with protein content. Selection for yield, agronomic characteristics and seed quality, large yellow seeds with typical averages of 20% oil and 40% protein were obtained. However, soybean is appreciated for its high protein meal and versatile vegetable oils; therefore, breeders mostly prefer to obtain modest gains in oil and yield without substantial loss in protein concentration [42]. Breeding for oil quality such as with reduced saturated fatty acids are prime focus as it is responsible for elevating cholesterol. The saturated fatty acids present in soybean oil are palmitic acid, 16:0, and stearic acid, 18:0. Especially, palmitic acid is a health concern as it is correlated to cardiovascular disease. It has been suggested that saturated fatty acids should be kept below 7–10% on a daily basis [42]. Soybean oil contains the monounsaturated fatty acid, oleic acid, 18:1. The oxidative stability of the soybean oil is enhanced by increasing three times higher the concentration of monounsaturated fatty acid such as oleic (18:1) than the normal content which is about 22%. Therefore, breeders target a concentration of 18:1 of about 65–75% of total lipid in soybean. By the means of genetic engineering, 18:1 levels of about 80% total lipid have been achieved [42]. In general, soybean varieties with unique fatty acid composition such as high oleic acid content, high stearic acid content, low linolenic acid content, or low palmitic acid content are preferred [13].

Assessment of agronomic traits has been used to evaluate phenotypic diversity in 20,570 Chinese soybean accessions and it was reported that seed coat color had the highest diversity index among the qualitative traits [126]. Plant's height had the most variation among quantitative traits, and followed by seed size, protein content, growth period, and oil content. The seed size of those accessions ranged from smaller than 2 to as large as 46 g/100-seeds. The protein content ranged from 30 to 53%; and oil content ranged from 10 to 25%. The variances of seed size, protein content, and oil content of the U.S. cultivars were lower than the Chinese cultivars [127]. The Southern U.S. soybeans were more variable in oil and protein contents and less variable in seed size than the Northern U.S. soybeans. The food-grade soybean breeding aims to increase the nutritional content and quality of protein and oil [128]. Greater genetic diversity of protein content, seed hardness, calcium content, and seed size uniformity than other quality traits in both small and large-seeded genotypes were evaluated [128]. The U.S. soybean genotypes with small seed were more diverse and exhibited higher swell ratio and oil content but lower stone seed ratio and protein content than the Asian accessions [128]. Among the large-seeded accessions, the U.S. genotypes had higher stone seed ratio and oil content but lower swell ratio and protein content, and were less diverse than the Asian genotypes [128]. The characterization of diverse food grade soybeans will facilitate parent selection in specialty soybean breeding [1].

12. Breeding for reduced trypsin inhibitor

Soybean germplasm PI542044, also known as Kunitz soybean, contains the null allele of KTI, i.e., kti that encodes a truncated protein and it was developed in a backcross program involving Williams 82 and PI157440 [129]. Introgression of kti is complicated by a number of factors viz., (i) kti being recessive in inheritance, each conventional backcross generation would be requiring selfing followed by estimation of KTI content in the seeds so as to identify a target plant. However, three recessive null alleles, viz. Kunitz trypsin inhibitor, soybean agglutinin, and P34 allergen null were stacked in the background of "Williams 82" and were termed as "Triple Null" [130]. Three SSR markers, viz. Satt228, Satt409, and Satt429 have been reported to be closely linked (0–10 cM) with the null allele of Kunitz trypsin inhibitor [131]. These SSR markers was also validated in the mapping population generated using Indian soybean genotypes as the recipient parent (*TiTi*) and PI542044 (*titi*) as the donor for the null allele [132]. Further, a gene-specific marker has also been designed from the null allele of KTI from genotype PI157440 [15] and has been deployed in the selection of plants carrying the null allele of KTI derived from PI542044 [121]. The null allele of KTI from PI542044 was introgressed into the cultivar "JS97–52" (recurrent parent) through marker-assisted backcrossing using the SSR marker Satt228, tightly linked with a trypsin inhibitor *Ti* locus. An introgressed line JS97–52 with reduced trypsin inhibitor (68.8–83.5%) content was developed [133].

13. Breeding for reduced cadmium content in soybean

Based on the importance of soybean as a staple food crop, the development of low Cd soybean cultivars should be a priority. The genetic variability for Cd accumulation within a species provides an opportunity to select soybean genotypes with low Cd concentration. In soybean grain, Cd concentration was found to be controlled by a single gene, with low Cd dominant in the crosses studied [134]. Lines with the low Cd trait had restricted root-to-shoot translocation, which limited the Cd accumulation in the grain. Genetic variability in soybean [19, 135] has been reported. An understanding of genetics and heritability of the Cd accumulation is essential in designing the breeding strategy to incorporate gene(s) controlling low Cd accumulation in modern cultivars. However, identifying low Cd phenotypes by analysis of the grain is challenging due to the high cost of analysis [136]. Developing inexpensive methods would assist in transferring the low Cd accumulation traits with other desirable traits.

14. Developing markers for marker-assisted selection of low Cd accumulation

Marker-assisted selection (MAS) could be an alternative to phenotypic selection. In soybean, DNA markers linked to low Cd accumulation were identified using RIL population (*F*6:8) derived from the cross AC Hime (high Cd accumulation in seeds) and Westag-97 (low Cd accumulation in seeds). The distribution of Cd concentration of 166 RILs ranged from 0.067 to 0.898 mg kg^{-1}, with a mean of 0.268 ± 0.013 mg kg^{-1} [134]. Using the RIL population, seven simple sequence repeat (SSR) markers, SatK138, SatK139, SatK140 (0.5 cM), SatK147, SacK149, SaatK150, and SattK152 (0.3 cM), were reported to be linked to *Cda1* in soybean seed. It was also reported that all the linked markers were mapped to the same linkage group (LG) K. SSR markers closely linked to *Cda*1 in soybean seeds have the potential to be used for MAS to develop low Cd-accumulating cultivars in a breeding program [134]. In a similar mapping approach, Benitez et al. [137] identified a major QTL cd1 on chromosome 9 (LG-K) across years and generations which accounted for 82, 57, and 75% of the genetic variation. Near-isogenic lines (NILs) were used to confirm the effect of the QTL and the peak of the QTL that was located in the vicinity of two SSR markers, Gm09:4770663 and Gm09:4790483. Both the studies revealed a major QTL for seed *Cd* content, *Cda*1 at a similar genomic location, suggesting that *cd*1 and *Cda1* may be identical. Candidate genes related to heavy metal transport or homeostasis were located in the vicinity of the identified QTL (Cda1). Protein kinase, putative adagio-like protein, and plasma membrane H^+-ATPase were found in the QTL vicinity. The presence of protein kinase and plasma membrane H^+-ATPase genes near the tightly linked SSR markers suggests that the regulation of this enzyme may play a vital role in Cd stress [134]. This was later supported by a major QTL-controlling Cd concentration (*cd*1) identified in soybean [137]. The gene was designated as *GmHMA1.* In *GmHMA1a*, one base substitution from G to A at nucleotide position 2095 resulted in a loss of function of the ATPase and was found to be associated with Cd uptake [137]. The SSR markers linked to the *Cda*1 and*Cd*1gene(s)/or QTLs and the SNP marker in the P1B-ATPase metal ion transporter gene in soybean can be utilized in MAS for developing soybean cultivars with low Cd content.

15. Future directions

Breeding for soybean seed composition traits is a complicated process; fortunately, ample genomic resources and tools are now available to soybean breeders/researchers for dissection of seed composition traits. The combination of conventional breeding strategy and genomic approaches will help to identify genomic loci, haplotypes, and FMs in breeding for improvement of seed composition traits. For improvement of protein, the major protein QTL, which was repeatedly mapped on Chr20, Chr15, and Chr18, may facilitate breeders to select parental lines and consider them for crossing schemes or introgression into locally adapted superior yielding cultivars through genomics-assisted breeding and MAS. Issues related to protein increase without yield drag, pleiotropic effects, and background/allelic effects could be addressed via screening diverse germplasm, considering wild soybean alleles for introgression, undertaking genomics-assisted breeding, precise high-throughput phenotyping, mutational breeding, and genome editing through Crisp/Cas. Integrating these aspects will extend our current genetic and genomic portfolio far beyond that of traditional breeding. Finally, when a cultivar with improved food-grade characteristics is developed, a further step is the evaluation of the quality of the product obtained from this cultivar. This is important as the success of a food-grade soybean cultivar is determined by the preferences of the consumers.

Author details

Souframanien Jegadeesan[1] and Kangfu Yu[2*]

1 Nuclear Agriculture and Biotechnology Division, Bhabha Atomic Research Centre, Mumbai, Maharashtra, India

2 Harrow Research and Development Centre, Agriculture and Agri-Food Canada, Ontario, Canada

*Address all correspondence to: kangfu.yu@canada.ca

References

[1] Zhang B, Chen P, Florez-Palacios SL, Shi A, Hou A, Ishibashi T. Seed quality attributes of food-grade soybeans from the U.S. and Asia. Euphytica. 2010;**173**:387-396

[2] Cui ZAT, James S, Miyazaki RF, Wilson TE. Carter. Breeding of specialty soybeans for traditional and new soyfoods. In: Liu K, editor. Soybeans as a Functional Food. Champaign: AOCS Press; 2004. pp. 264-322

[3] Omoni AO, Aluko RE. Soybean foods and their benefits: Potential mechanisms of action. Nutrition Reviews. 2005;**63**:272-283

[4] US Soyfoods Directory. 2006. Available from: http://www.soyfoods.com/soyfoodsdescriptions/descriptions.html

[5] Soyatech. Soyfood Facts. Available from: http://www.soyatech.com/soyfoods_facts.htm [Cited: 20 January 2012]

[6] L'Hocine L, Boye JI. Allergenicity of soybean: New developments in identification of allergenic proteins, cross-reactivities and hypo-allergenization technologies. Critical Reviews in Food Science and Nutrition. 2007;**47**:127-143

[7] Jaureguy LM, Chen P, Scaboo AM. Heritability and correlations among food-grade traits in soybean. Plant Breeding. 2011;**130**:647-652

[8] Graef GL, Specht JE. Fitting the Niche Food Grade Soybean Production: A New Opportunity for Nebraska Soybean Producers. Lincoln: Nebraska Department of Agriculture; 1989. pp. 18-27

[9] Poysa V, Cober E, Reid J, Pietrzak L, Mullin J, Voldeng H. Improving the Quality of Food Grade Soybeans. Bulletin National Research Council (NRC), Plant Biology Institute (PBI). 2002

[10] Bachman MS, Tamulonis JP, Nickell CD, Bent AF. Molecular markers linked to brown stem rot resistance genes, Rbs1 and Rbs2, in soybean. Crop Science. 2001;**41**:527-535

[11] Mullin WJ, Xu W. Study of soybean seed coat components and their relationship to water absorption. Journal of Agricultural and Food Chemistry. 2001;**49**:5331-5335

[12] Saio K. Soybeans resistant to water absorption. Cereal Foods World. 1976;**21**:168-173

[13] Chen P. Developing high quality identity-preserved soybean for the specialty soyfood market. In: Production and Marketing of Identity-Preserved Soybean. MO, USA: American Soybean Association; 2004. pp. 23-31

[14] Taira H. Quality of soybeans for processed foods in Japan. Japan Agricultural Research Quarterly. 1990;**24**:224-230

[15] Moraes R, Soares CB, Colombo LR, Salla MFS, Barros JGA, Piovesan ND, et al. Assisted selection by specific DNA markers for genetic elimination of the kunitz trypsin inhibitor and lectin in soybean seeds. Euphytica. 2006;**149**:221-226

[16] Raboy V. Seed phosphorous and the development of low-phytate crop. In: Inositol Phosphates: Linking Agriculture and the Environment Turner BL, Richardson AE, Mullany EJ.). Wallingford, U.K: CAB International; 2007. pp. 111-132.

[17] Talukdar A, Shivkumar M. Genetic improvement of food-grade soybean in India: Current status and future prospects. The Indian Journal of Genetics. 2016;**76**:626-630

[18] McGrath SP, Zhao FJ, Lombi E. Plant and rhizosphere processes involved

in phyto remediation of metal-contaminated soils. Plant and Soil. 2001;**232**:207-214

[19] Arao T, Ae N, Sugiyama M, Takahash M. Genotypic differences in cadmium uptake and distribution in soybeans. Plant and Soil. 2003;**251**:247-253

[20] Sugiyama M, Ae N, Hajika M. Developing of a simple method for screening soybean seedling cadmium accumulation to select soybean genotypes with low seed cadmium. Plant and Soil. 2011;**341**:413-422

[21] Kazantzis G. Cadmium, osteoporosis and calcium metabolism. Biometals. 2004;**17**:493-498

[22] Codex Alimentarius Commission. Report of the 33rd Session of the Codex Committee on Food Additives and Contaminants. Hague, The Netherlands: Codex Alimentarius Commission; 2001. p. 285

[23] Ministry of Agriculture, Forestry, and Fisheries of Japan. Investigation of Cd Concentration of Staple Crops: The Outline of the Results. Tokyo: Ministry of Agriculture, Forestry, and Fisheries of Japan; 2002

[24] Cupit M, Larsson O, de Meeus C, Eduljee GH, Hutton M. Assessment and management of risks arising from exposure to cadmium in fertilisers: II. Science of the Total Environment. 2002;**291**:189-206

[25] Endres JG. Protein quality and human nutrition. In: Endres JG, editor. Soy Protein Products: Characteristics, Nutritional Aspects, and Utilization. Revised and Expanded Edition. Champaign, Illinois, USA: AOCS; 2001. pp. 10-19

[26] Kumar V, Rani A, Chauhan GS. Nutritional value of soybean. In: Singh G, editor. The Soybean: Botany, Production and Uses. Wallingford, U.K: CABI; 2010. pp. 375-403

[27] Liu K. Soybeans as a powerhouse of nutrients and phytochemicals. In: Liu K, editor. Soybeans as Functional Foods and Ingredients. New York: AOCS Publishing; 2004

[28] Poysa V, Woodrow L. Stability of soybean seed composition and its effect on soymilk and tofu yield and quality. Food Research International. 2002;**35**:337-345

[29] Cai T, Chang KC. Processing effect on soybean storage proteins and their relationship with tofu quality. Journal of Agricultural and Food Chemistry. 1999;**47**:720-727

[30] Poysa V, Woodrow L, Yu K. Effect of soy protein subunit composition on tofu quality. Food Research International. 2006;**39**:309-317

[31] Nielsen NC, Dickinson CD, Cho TJ, Thanh VH, Scallon BJ, Fischer RL, et al. Characterization of the glycinin gene family in soybean. The Plant Cell. 1989;**1**:313-328

[32] Maruyama N, Adachi M, Takahashi K, Yagasaki K, Kohna M, Takenaka Y, et al. Crystal structures of recombinant and native soybean β -conglycinin β homotrimers. European Journal of Biochemistry. 2001;**268**:3595-3604

[33] Thanh VH, Shibasaki KK. Major proteins of soybean seeds. A straight forward fractionation and their characterization. Journal of Agricultural and Food Chemistry. 1976;**24**:1117-1121

[34] Ogawa T, Bando N, Tsuji H, Nishikawa K, Kitamura K. 1995. Alpha-subunit of beta-conglycinin, an allergenic protein recognized by IgE antibodies of soybean-sensitive patients with atopic dermatitis. Bioscience,

Biotechnology, and Biochemistry. 59:831-833.

[35] Panthee DR, Kwanyuen P, Sams CE, West DR, Saxton AM, Pantalone VR. Quantitative trait loci for β-conglycinin (7S) and glycinin (11S) fractions of soybean storage protein. Journal of the American Oil Chemists' Society. 2004;**81**:1005-1012

[36] Yu K, Woodrow L, Shi C, Anderson D, Poysa V. Registration of 7S β-conglycinin α′ and 11S glycinin A4 null food-grade soybean germplasm, HS-162. Canadian Journal of Plant Science. 2017;**97**:536-538

[37] Yagasaki K, Kousaka F, Kitamura K. Potential improvement of soymilk gelation properties using soybean with modified protein subunit composition. Breeding Science. 2000;**50**:101-107

[38] Tezuka SL, Taira H, Igarashi Y, Yagasaki K, Ono T. Properties of tofus and soy milks prepared from soybeans having different subunits of glycinin. Journal of Agricultural and Food Chemistry. 2000;**48**:1111-1117

[39] Nik AM, Tosh SM, Woodrow L, Poysa V, Corredig M. Effect of soy protein subunit composition and processing conditions on stability and particle size distribution of soymilk. LWT- Food Science and Technology. 2009;**42**:1245-1252

[40] Yu K, Woodrow L, Poysa V. Registration of lipoxygenase free food grade soybean Germplasm, HS-151. Canadian Journal of Plant Science. 2016;**96**:148-150

[41] Obendorf RL, Kosina SM. Soluble carbohydrates in soybean. In: Tzi-Bun N, editor. Soybean-Biochemistry, Chemistry and Physiology. UK: Intechopen; 2011. pp. 201-228

[42] Cober ER, Cianzion SR, Pantalone VR, Rajcan I. In: Vollmann J, Rajcan I, editors. Soybean, in Oil Crops. Handbook of Plant Breeding. Vol. 4. New York; Springer; 2009. pp. 57-90

[43] Sathyapalan T, Aye M, Rigby A, Thatcher NJ, Dargham SR, Kilpatrick ES, et al. Soy isoflavones improve cardiovascular disease risk markers in women during the early menopause. Nutrition, Metabolism, and Cardiovascular Diseases. 2018;**28**:691-697

[44] Esler I. Prospects for Vegetable Soybean in India and Its Market Acceptance. Report. Taiwan: AVRDC—The World Vegetable Centre; 2011. pp. 1-26

[45] Lee JY, Min S, Choe EO, Min DB. Formation of volatile compounds in soy flour by singlet oxygen oxidation during storage under light. Journal of Food Science. 2003;**68**:1933-1937

[46] Iassonova DR, Johnson LA, Hammond EG, Beattie SE. Evidence of anenzymatic source of off flavors in lipoxygenase-null soybeans. Journal of the American Oil Chemists' Society. 2009;**86**:59-64

[47] Buzzell RI, Anderson TR, Hamill AS, Welacky TW. Harovinton soybean. Canadian Journal of Plant Science. 1991;**71**:525-526

[48] Griffis G, Wiedermann L. Marketing Food-Quality Soybeans in Japan. 3rd ed. St. Louis: American Soybean Association; 1990

[49] Wilson RF. Seed composition. In: Boerma HR, Specht JE, editors. Soybeans: Improvement, Production, and Uses. 3rd ed. Madison: American Society of Agronomy, Inc./Crop Science Society of America, Inc./Soil Science Society of America, Inc.; 2004, 2004. p. 621

[50] Hwang E, Song Q, Jia G, Specht J, Hyten D, Costa J. A genome-wide association study of seed protein

and oil content in soybean. BMC Genomics. 2014;**15**:1

[51] Eskandari M, Cober ER, Rajcan I. Genetic control of soybean seed oil: II. QTL and genes that increase oil concentration without decreasing protein or with increased seed yield. Theoretical and Applied Genetics. 2013;**126**:1677-1687

[52] SoyBase. Available from: https://soybase.org/

[53] Pathan SM, Vuong T, Clark K, Lee JD, Shannon JG, Roberts CA, et al. Genetic mapping and confirmation of quantitative trait loci for seed protein and oil contents and seed weight in soybean. Crop Science. 2013;**53**:765-774

[54] Akond M, Ragin B, Bazzelle R, Kantartzi SK, Meksem K. Quantitative trait loci associated with moisture, protein, and oil content in soybean [*Glycine max* (L.) Merr.]. Journal of Agricultural Science. 2012;**4**:16-25

[55] Diers B, Keim P, Fehr W, Shoemaker R. RFLP analysis of soybean seed protein and oil content. Theoretical and Applied Genetics. 1992;**83**:608-612

[56] Nichols D, Glover K, Carlson S, Specht J, Diers B. Fine mapping of a seed protein QTL on soybean linkage group I and its correlated effects on agronomic traits. Crop Science. 2006;**46**:834-839

[57] Brummer EC, Graef GL, Orf J, Wilcox JR, Shoemaker RC. Mapping QTL for seed protein and oil content in eight soybean populations. Crop Science. 1997;**37**:370

[58] Lee SH, Bailey MA, Mian MAR, Carter TE, Shipe ER, Ashley DA, et al. RFLP loci associated with soybean seed protein and oil content across populations and locations. Theoretical and Applied Genetics. 1996;**93**:649-657

[59] Sebolt A, Shoemaker R, Diers B. Analysis of a quantitative trait locus allele from wild soybean that increases seed protein concentration in soybean. Crop Science. 2000;**40**:1438-1444

[60] Chung J, Babka H, Graef G, Staswick P, Lee D, Cregan P, et al. The seed protein, oil, and yield QTL on soybean linkage group I. Crop Science. 2003;**43**:1053-1067

[61] Fasoula VA, Harris DK, Boerma HR. Validation and designation of quantitative trait loci for seed protein, seed oil, and seed weight from two soybean populations. Crop Science. 2004;**44**(4):1218

[62] Panthee DR, Pantalone VR, West DR, Saxton AM, Sams CE. Quantitative trait loci for seed protein and oil concentration, and seed size in soybean. Crop Science. 2005;**45**:2015

[63] Lu W, Wen Z, Li H, Yuan D, Li J, Zhang H, et al. Identification of the quantitative trait loci (QTL) underlying water soluble protein content in soybean. Theoretical and Applied Genetics. 2013;**126**:425-433

[64] Wang J, Chen P, Wang D, Shannon G, Zeng A, Orazaly M, et al. Identification and mapping of stable QTL for protein content in soybean seeds. Molecular Breeding. 2015;**35**:1-10

[65] Phansak P, Soonsuwon W, Hyten DL, Song Q, Cregan PB, Graef GL, et al. Multi-population selective genotyping to identify soybean (*Glycine max* (L.) Merr.) seed protein and oil QTLs. G3: Genes, Genomes, Genetics. 2016;**6**:1635-1648

[66] Vaughn JN, Nelson RL, Song Q, Cregan PB, Li Z. The genetic architecture of seed composition in soybean is refined by genome-wide association scans across multiple populations. G3: Genes, Genomes, Genetics. 2014;**4**:2283-2294

[67] Sonah H, O'Donoughue L, Cober E, Rajcan I, Belzile F. Identification of loci governing eight agronomic traits using a GBSGWAS approach and validation by QTL mapping in soybean. Plant Biotechnology Journal. 2015;**3**:211-221

[68] Bandillo N, Jarquin D, Song QJ, Nelson R, Cregan P, Specht J, et al. A population structure and genome-wide association analysis on the USDA soybean germplasm collection. Plant Genome. 2015;**8**:1-13

[69] Valliyodan B, Qiu D, Patil G, Zeng P, Huang J, Dai L, et al. Landscape of genomic diversity and trait discovery in soybean. Scientific Reports. 2016;**6**:23598

[70] Zhou Z, Jiang Y, Wang Z, Gou Z, Lyu J, Li W, et al. Resequencing 302 wild and cultivated accessions identifies genes related to domestication and improvement in soybean. Nature Biotechnology. 2015;**33**:408-414

[71] Tajuddin T, Watanabe S, Yamanaka N, Harada K. Analysis of quantitative trait loci for protein and lipid contents in soybean seeds using recombinant inbred lines. Breeding Science. 2003;**53**:133-140

[72] Hyten DL, Pantalone VR, Sams CE, Saxton AM, Landau-Ellis D, Stefaniak TR, et al. Seed quality QTL in a prominent soybean population. Theoretical and Applied Genetics. 2004;**109**:552-561

[73] Maughan PJ, Maroof MAS, Buss GR. Identification of quantitative trait loci controlling sucrose content in soybean (*Glycine max* L.). Molecular Breeding. 2000;**6**:105-111

[74] Kim HK, Kang ST, Cho JH, Choung MG, Suh DY. Quantitative trait loci associated with Oligo saccharide and sucrose contents in soybean (*Glycine max* L.). Journal of Plant Biology. 2005;**48**:106-112

[75] Kim HK, Kang ST, Oh DW. Mapping of putative quantitative trait loci controlling the total oligosaccharide and sucrose content of *Glycine max* seeds. Journal of Plant Research. 2006;**119**:533-538

[76] Akond M, Liu S, Kantartzi SK, Meksem K, Bellaloui N, David, et al. Quantitative trait loci underlying seed sugars content in "MD96-5722" by "Spencer" recombinant inbredline population of soybean. Food and Nutrition Sciences. 2015;**6**:964-973

[77] Cicek M. Genetic marker analysis of three major carbohydrates in soybean seeds [PhD dissertation]. Blacksburg: Virginia Polytechnic Institute & State University; 2001

[78] Clevinger EM. Mapping quantitative trait loci for soybean quality traits from two different sources [PhD dissertation]. Blacksburg: Virginia Polytechnic Institute and State University; 2006

[79] Maroof MAS, Buss GR. Low phytic, low stachyose, high sucrose soybean lines. Patent Application Number: 20080199591; US20080199591 A1; US 8003856 B2; 2008

[80] Feng C, Morsy M, Giannoccaro E, Zhang B, Chen P. Soybean seed sugar content and quantitative trait loci mapping. In: Li CJ et al., editors. Plant Nutrition for Food Security, Human Health, and Environmental Protection. Beijing: Tsinghua University Press; 2005. p. 438

[81] Zeng A, Chen P, Shi A, Wang D, Zhang B, Orazaly M, et al. Identification of quantitative trait loci for sucrose content in soybean seed. Crop Science. 2014;**54**:554-564

[82] Wang Y, Chen P, Zhang B. Quantitative trait loci analysis of soluble sugar contents in soybean. Plant Breeding. 2014;**133**:493-498

[83] Patil G, Chaudhary J, Vuong TD, Jenkins B, Qiu D, Kadam S, et al. Development of SNP genotyping assays for seed composition traits in soybean. International Journal of Plant Genomics. 2017:1-12. DOI: 10.1155/2017/6572969

[84] Poeta F, Borrás L, Rotundo JL. Variation in seed protein concentration and seed size affects soybean crop growth and development. Crop Science. 2016;**6**:3196-3208

[85] Ishikawa G, Takada Y, Nakamura T. A PCR based method to test for the presence or absence of β-conglycinin α′- and α-subunits in soybean. Molecular Breeding. 2006;**17**:365-374

[86] Tsubokura Y, Hajika M, Harada K. Molecular marker associated with β –conglycinin deficiency in soybean. Breeding Science. 2006;**56**:113-117

[87] Teraishi M, Takahashi M, Hajika M, Matsunaga R, Uematsu Y, Ishimoto M. Suppression of soybean β-conglycinin genes by a dominant gene, Scg-1. Theoretical and Applied Genetics. 2001;**103**:1266-1272

[88] Hayashi M, Nishioka M, Kitamura K, Harda K. Identification of AFLP markers tightly linked to the gene for deficiency of the 7S globulin in soybean seed and characterization of abnormal phenotypes involved in the mutation. Breeding Science. 2000;**50**:123-129

[89] Diers BW, Beilinson V, Nielsen NC, Shoemaker RC. Genetic mapping of the Gy4 and Gy5 glycinin genes in soybean and the analysis of a variant of Gy4. Theoretical and Applied Genetics. 1994;**89**:297-304

[90] Beilinson V, Chen Z, Shoemaker RC, Fischer RL, Goldberg RB, Neilsen NC. Genomic organization of glycinin genes in soybean. Theoretical and Applied Genetics. 2002;**104**:1132-1140

[91] Boehm JD Jr, Nguyen V, Tashiro RM, Anderson D, Shi C, Wu X, et al. Genetic mapping and validation of the 7S α' and 11S A-type storage protein subunits in soybean [*Glycine max* (L.) Merr.]. Theoretical and Applied Genetics (TAG). 2018;**131**(3):659-671

[92] Andersen JR, Lubberstedt T. Functional markers in plants. Trends in Plant Science. 2003;**8**:554-560

[93] Wang C, Wu X, Jia F, Zhang J, Chen S. Genetic variations of glycinin genes among cultivated and wild type soybean species. Progress in Natural Science. 2008;**18**:33-41

[94] Mahmoud AA, Natarajan SS, Bennett JO. Effect of six decades of selective breeding on soybean protein composition and quality: A biochemical and molecular analysis. Journal of Agricultural and Food Chemistry. 2006;**54**:3916-3922

[95] Souframanien J, Yu K, Lorna Woodrow L, Wang Y, Shi C, Poysa V. Molecular analysis of glycinin genes in soybean mutants for development of gene specific markers. Theoretical and Applied Genetics. 2012;**124**:365-372

[96] Mandal S, Mandal R. Seed storage proteins and approaches for improvement of their nutritional quality by genetic engineering. Current Science. 2000;**79**:576-589

[97] Warrington C, Abdel-Haleem H, Hyten D, Cregan P, Orf J, Killam A, et al. QTL for seed protein and amino acids in the Benning × Danbaekkong soybean population. Theoretical and Applied Genetics. 2015;**128**:839-850

[98] Ma Y, Kan G, Zhang X, Wang Y, Zhang W, Du H, et al. Quantitative trait loci (QTL) mapping for glycinin and beta-conglycinin contents in soybean (*Glycine max* L. Merr.). Journal of Agricultural and Food Chemistry. 2016;**64**:3473-3483

[99] Panthee D, Pantalone V, Sams C, Saxton A, West D, Orf J, et al. Quantitative trait loci controlling sulfur containing amino acids, methionine and cysteine, in soybean seeds. Theoretical and Applied Genetics. 2006;**112**:546-553

[100] Chaudhary J, Patil GB, Sonah H, Deshmukh RK, Vuong TD, Valliyodan B, et al. Expanding omics resources for improvement of soybean seed composition traits. Frontiers in Plant Science. 2015;**6**:1021

[101] Krishnan HB, Kim WS, Oehrle NW, Alaswad AA, Baxter I, Wiebold WJ, et al. Introgression of leginsulin, a cysteine-rich protein, and high-protein trait from an Asian soybean plant introduction genotype into a North American experimental soybean line. Journal of Agricultural and Food Chemistry. 2015;**63**:2862-2869

[102] Deshmukh R, Sonah H, Patil G, Chen W, Prince S, Mutava R. Integrating omic approaches for abiotic stress tolerance in soybean. Frontiers in Plant Science. 2014;**5**:244

[103] Yates JL. Use of diverse germplasm to improve peanut root-knot nematode resistance and seed protein content in soybean [PhD dissertation]. Athens: University of Georgia; 2006

[104] Mian MA, McHale L, Li Z, Dorrance AE. Registration of 'Highpro1' soybean with high protein and high yield developed from a north × south cross. Journal of Plant Registrations. 2017;**11**:51-54

[105] Bolon Y-T, Stec AO, Michno J-M, Roessler J, Bhaskar PB, Ries L, et al. Genome resilience and prevalence of segmental duplications following fast neutron irradiation of soybean. Genetics. 2014;**198**:967-981

[106] Sandhu D, Alt JL, Scherder CW, Fehr WR, Bhattacharyya MK. Enhanced oleic acid content in the soybean mutant M23 is associated with the deletion in the Fad2-1a gene encoding a fatty acid desaturase. Journal of the American Oil Chemists' Society. 2007;**84**:229-235

[107] Schmidt MA, Herman EM. Suppression of soybean oleosin produces micro-oil bodies that aggregate into oil body/ER complexes. Molecular Plant. 2008;**1**:910-924

[108] Gillman JD, Stacey MG, Cui Y, Berg HR, Stacey G. Deletions of the SACPD-C locus elevate seed stearic acid levels but also result in fatty acid and morphological alterations in nitrogen fixing nodules. BMC Plant Biology. 2014;**14**:1

[109] Kitamura K. Genetic improvement of nutritional and food processing quality in soybean. JARQ. 1995;**29**:1-8

[110] Song B, Shen L, Wei X, Guo B, Tuo Y, Tian F, et al. Marker-assisted backcrossing of a null allele of the α-subunit of soybean (*Glycine max*) β-conglycinin with a Chinese soybean cultivar (a). The development of improved lines. Plant Breeding. 2014;**133**:638-648

[111] Brim C, Burton J. Recurrent selection in soybeans. II. Selection for increased percent protein in seeds. Crop Science. 1979;**19**:494-498

[112] Holbrook CC, Burton JW, Carter TE. Evaluation of recurrent restricted index selection for increasing yield while holding seed protein constant in soybean. Crop Science. 1989;**29**:324-329

[113] Lewers K, Palmer R. Recurrent selection in soybean. Plant Breeding Reviews. 1997;**15**:275-314

[114] Varshney RK, Mohan SM, Gaur PM, Gangarao N, Pandey MK, Bohra A, et al. Achievements and prospects of genomics-assisted breeding

in three legume crops of the semi-arid tropics. Biotechnology Advances. 2013;**31**:1120-1134

[115] Jarquin D, Specht J, Lorenz A. Prospects of genomic prediction in the USDA soybean germplasm collection: Historical data creates robust models for enhancing selection of accessions. G3: Genes, Genomes, Genetics. 2016;**116**:031443

[116] Patil G, Do T, Vuong TD, Valliyodan B, Lee JD, Chaudhary J, et al. Genomic-assisted haplotype analysis and the development of high-throughput SNP markers for salinity tolerance in soybean. Scientific Reports. 2016;**6**:19199

[117] Tanksley S, Nelson J. Advanced backcross QTL analysis: A method for the simultaneous discovery and transfer of valuable QTLs from unadapted germplasm into elite breeding lines. Theoretical and Applied Genetics. 1996;**92**:191-203

[118] Mozzoni L, Shi AN, Chen PY. Genetic analysis of high sucrose, low Raffinose, and low Stachyose content in V99-5089 soybean seeds. Journal of Crop Improvement. 2013;**27**:606-616

[119] Axelord B, Cheesborough TM, Laasko S. Lipoxygenase from soybean. Methods in Enzymology. 1981;**71**:441

[120] Xu SJ, Singh RJ, Kalipara KP, Hymowitz T. Primary trisomics in soybean: Origin, identification, breeding behaviour and use in linkage mapping. Crop Science. 2000;**40**:1543

[121] Kumar V, Rani A, Rawal R, Hussain SM. Lipoxygenase-2 free Indian soybean genotype. Current Science. 2013;**104**:586

[122] Reinprecht Y, Luke-Labey SY, Yu K, Poysa VW, Rajcan I, Ablett GR, et al. Molecular basis of seed lipoxygenase-null traits in soybean line OX948. Theoretical and Applied Genetics. 2011;**122**:1247

[123] Gutierrez-Gonzalez JJ, Vuong TD, Zhong R, Yu O, Lee JD, Shannon G, et al. Major locus and other novel additive and epistatic loci involved in modulation of isoflavone concentration in soybean seeds. Theoretical and Applied Genetics. 2011;**123**:1375-1385

[124] Li B, Tian L, Zhang J, Huang L, Han F, Yan S, et al. Construction of a high-density genetic map based on large-scale markers developed by specific length amplified fragment sequencing (SLAF-seq) and its application to QTL analysis for isoflavone content in *Glycine max*. BMC Genomics. 2014;**15**:1086

[125] Akond M, Liu S, Kantartzi SK, Meksem K, Bellaloui N, Lightfoot DA, et al. Quantitative trait loci for seed Isoflavone contents in "MD96-5722" by "Spencer" recombinant inbred lines of soybean. Journal of Agricultural and Food Chemistry. 2014;**62**:1464-1468

[126] Dong YS, Zhao LM, Liu B, Wang ZW, Jin ZQ, Sun H. The genetic diversity of cultivated soybean grown in China. Theoretical and Applied Genetics. 2004;**108**:931-936

[127] Cui Z, Carter TE, Burton JW, Wells R. Phenotypic diversity of modern Chinese and north American soybean cultivars. Crop Science. 2001;**41**:1954-1967

[128] Liu K. Soybeans: Chemistry, Technology and Utilization. New York: Chapman & Hall; 1997. p. 523

[129] Bernard RL, Hymowitz T, Cremeens CR. Registration of "Kunitz" soybean. Crop Science. 1991;**31**:232-233

[130] Schmidt MA, Hymowitz T, Herman EM. Breeding and characterization of soybean triple null; a stack of recessive alleles of Kunitz trypsin inhibitor, soybean agglutinin, and P34 allergen nulls. Plant Breeding. 2015;**134**(3):310-315

[131] Kim MS, Park MJ, Jeong WH, Nam KC, Chung JI. SSR marker tightly linked to the Ti locus in soybean [*Glycine max* (L.) Merr.]. Euphytica. 2006;**152**:361-366

[132] Rani A, Kumar V, Mourya V, Singh RK, Husain SM. Validation of SSR markers linked to null kunitz trypsin inhibitor allele in Indian soybean [*Glycine max* (L.) Merr.] population. Journal of Plant Biochemistry and Biotechnology. 2011;**20**:258-261

[133] Kumar V, Rani A, Rawal R, Mourya V. Marker assisted accelerated introgression of null allele of kunitz trypsin inhibitor in soybean. Breeding Science. 2015;**65**:447-452

[134] Jegadeesan S, Yu KF, Poysa V, Gawalko E, Morrison MJ, Shi C, et al. Mapping andvalidation of simple sequence repeat markers linked to a major gene controlling seed cadmium accumulation in soybean [*Glycine max* (L.) Merr]. Theoretical and Applied Genetics. 2010;**121**:283-294

[135] Ishikawa S, Ae N, Sugiyama M, Murakami M, Arao T. Genotypic variation in shoot cadmium concentration in rice and soybean in soils with different levels of cadmium contamination. Soil Science & Plant Nutrition. 2005;**51**:101-108

[136] Grant CA, Clarke JM, Duguid S, Chaney RL. Selection and breeding of plant cultivars to minimize cadmium accumulation. Science of the Total Environment. 2008;**390**:301-310

[137] Benitez ER, Hajika M, Takahshi R. Single-base substitution in P1B-ATPase. Gene is associated with a major QTL for seed cadmium concentration in soybean. Journal of Heredity. 2012;**103**:278-286

Chapter 5

Organic Grain Legumes in India: Potential Production Strategies, Perspective, and Relevance

Amanpreet Singh and Harmandeep Singh Chahal

Abstract

Organic agriculture comes from the conscious efforts of people who are inspired to create the best possible relationship between the earth and humans. Since its inception, the sphere surrounding organic farming has become much more complex. The introduction of legumes into crop rotation is not a new concept, legumes such as peas and chickpeas were introduced into cereal-related crop rotation during the Harappan period of the chalcolithic period at the end of 3000 BC. The grain yield after legumes in the grain system is 30 to 35% higher than the grain in the crop chain. Legumes play a unique role in organic farming systems because of their deep root system, ability to fix nitrogen, and the ability to rotate and rotate systems.

Keywords: organic farming, legumes, nutrient, pest, constraints

1. Introduction

Organic agriculture comes from the conscious efforts of people who are inspired to create the best possible relationship between the earth and humans. Since its inception, the sphere surrounding organic farming has become much more complex. A major challenge of today's agriculture scenario is food contamination of unhealthy chemicals related to fertilizers, herbicides, and pesticides. Over the past two decades, the global community has also been sensitive to preserve the environment and food quality. Enthusiastic promoters of organic farming believe that it can meet these two requirements and become a comprehensive means of rural development. After nearly a century of development, organic farming has gained public acceptance and shows great commercial, social, and environmental promise. Although there has been a constant mindset from the very first days to the present day, the modern organic movement is completely different from its original form and based on environmental sustainability, as well as the founder's interest in healthy soil, healthy food, and healthy people. Organic farming is the best way to improve soil quality and the health of organisms. Today, the world recognizes the importance of quality food products that are not contaminated by synthetic materials, namely chemical fertilizers, pesticides and pesticides in agricultural production, and hormones and chemicals used in the livestock industry. Synthetic fertilizers and pesticides are not necessary to maintain a sufficient supply of nutritious food to feed the developing world population and worse, can lead to environmental degradation in general, and soil quality in particular. So manipulation of organic farms with the utilization of natural resources is necessary for the form of sustainable management based on the agronomic alteration of present agriculture scenario. Legumes have good potential to diversify farming systems on organic farms. Introducing legumes into organic production systems add to improving soil fertility by fixing nitrogen and releasing phosphorus nutrients. Therefore, in legume rotation, the subsequent grain yield and the crude protein concentration in the seed may increase due to excess nitrogen provided by the previous legume. Legumes help in increasing organic matter and microbial biomass, soil activity, improve soil structure and water holding capacity while helping to reduce water erosion. Crop rotation plays an important role in organic farming, and rotations include legumes that allow for future production in the same land. The versatility of legumes allows them to be used in biological systems in different ways: crop rotation, intercropping, level cultivation, and cover crops. To reach the full yield potential of crops in organic production systems at all times, legumes must account for at least 30% of the total crop area [1]. The introduction of legumes into crop rotation is not a new concept, legumes such as peas and chickpeas were introduced into cereal-related crop rotation during the Harappan period of the chalcolithic period at the end of 3000 BC. The grain yield after legumes in the grain system is 30–35% higher than the grain in the crop chain. Legumes play a unique role in organic farming systems because of their deep root system, ability to fix nitrogen, and the ability to rotate and rotate systems. The choice of the utilization of grain legumes in organic farming increases day by day due to the increasing consumption of organic productions due to improvement in organic markets.

2. Status of grain legumes under organic farming vs. conventional farming

Recognizing the importance of clean food for people, the demand for organic products, especially in developed countries, is increasing day by day. Worldwide, 1.8 million farm households in 162 countries practice organic farming on 37 million hectares of land. The largest areas of organically managed agricultural land are Oceania (12.1 million hectares or 33% of the world's organic agricultural land), Europe (10.6 million hectares or 29% organic agricultural land of the world), and Latin America (6.8 million ha or 23%). India now ranks 10th among the top 10 countries in the world for organic certified farmlands. The certified area covers 15% of the cultivated area with 0.72 million ha and the remaining 85% (3.99 million ha) is left in the forest and wilderness to collect small forest products. The total area under organic certification is 4.72 million ha (2013–2014). In terms of production, India has produced about 1.24 million tons of certified organic products including sugarcane, cotton, oilseeds, basmati rice, legumes, spices, tea, fruits, dried fruits, vegetables, coffee, and value-added products and organic products is not limited to food items but it also includes an organic cotton and other daily items, etc. Of all the states, Madhya Pradesh occupies the largest area of organic produce certification, followed by Himachal Pradesh and Rajasthan [2].

In India, the main grain legumes are based on pulses and oilseeds and the main crops are lentils, grams, peanuts, soybeans, and pigeons (**Table 1**), but the organic production of organic grain legumes is only 1.44 million tons [3]. Organic production of grain legumes is low as compared to conventionally produced legumes but their production was improved day by day by the implementation of new research modules developed by agricultural institutes. Organic grain legumes not in terms of consumption are better utilized but it is developed as a backbone of agriculture markets in terms of export to other countries at higher rates. Soybeans (70%)

Crop	Total area (MH)	Total production (MT)
Lentil	1.55	1.61
Gram	10.56	11.23
Groundnut	4.91	9.18
Soybean	10.47	10.98
Pigeonpea	4.43	4.25

Agricultural Statistics at a Glance 2018 (Government of India).

Table 1.
Major grain legumes in India.

topped export products, followed by cereals and millet different from basmati (6%), processed food products (5%), basmati rice (4%), sugar (3%), tea (2%), legumes (1%), nuts (1%), spices (1%) and others (www.opeda.org). The Indian Government launched the National Organic Production Program (NPOP) in 2001. States like Uttarakhand, Karnataka, Madhya Pradesh, Maharashtra, Gujarat, Rajasthan, Tamil Nadu, Kerala, Nagaland, Mizoram, and Sikkim have encouraged organic agriculture based on cereals, pulses, oilseeds, and horticultural crops.

3. Crop production techniques in organic grain legumes

3.1 Choose the right variety

The legume improvement program began in 1917 with selections from different parts of the country, especially pigeonpea. Today, a large number of improved varieties have been published to improve yield, resistance to pests and pests, short duration, synchronized maturation, a small size, etc., adapting to conditions of different farmland. Generally, genetically improved varieties are not allowed in organic farms so selection or crop improvement in legume varieties should be based on adaptability to the agro-climatic conditions of locations and resistant to abiotic and biotic stress should be developed. The old desi varieties have low productivity, but their potential towards stress conditions was so good and acceptable.

3.2 Sowing time

Planting time is the most important non-monetary contribution that significantly affects crop growth and productivity [4]. A delay in planting limits the vegetative growth; a number of fruit-bearing branches or stems branches and leads to forced maturity and poor productivity. At the same time, the incidence of pests, especially the European stem borer (*Helicoverpa armigera*) on green beans [5]. Growing lentils from November 15th to 20th in Tripura gives maximum grain yield compared to early or late planting [6]. Agronomic alteration based on sowing time or date of sowing is a good manipulation on organic farms without the utilization of any unnecessary source.

3.3 Water management

In India's, there are two main water management issues: (i) water surplus and subsequent flooding problem during the Kharif season and (ii) water shortage during the Rabi season (November to April). Legumes need adequate drainage because they are prone to water-logging [7]. Legumes are mainly grown as rainfed crops in the semi-arid and arid areas. However, legumes grown during the summer months require significant irrigation, especially during the critical period, when soil moisture becomes a limiting factor. The vegetative, flowering and fruiting stages are very important in pulse physiology so adequate management of irrigation scheduling at these stages is important. In water deficit areas, alternate-furrow irrigation and water conservation techniques is a good approach.

3.4 Tillage

Arable land is necessary to treat the soil with agricultural tools and tools to obtain ideal conditions for seed germination, seed establishment, and plant growth. The main objective of tillage is to create good soil and soil conditions for crop establishment and initial development of roots and shoots. The Kharif legume requires tillage to open the soil through a rotating plow and two cross wrinkles followed by a plank. In Rabi legumes, soil plows after Kharif and if necessary, irrigation should be given before planting to ensure adequate moisture. Recycling of summer mung bean residues (*Vigna radiata*) through rotary tillers and cropping systems increases system productivity, profitability, and soil health [8]. Crop residue incorporation in organic fields with the help of tillage techniques is also a good approach for management of fertilizer needs of crops and improves organic matter content of the soil.

3.5 Geometric planting

The optimal space needed depends on the type of crop, the variety, the planting season, and the cropping system. Most short-lived legumes need a small space, while long-term varieties work well with a larger space. Appropriate planting densities in fields and vegetables lead to better use of solar radiation to turn into higher yields. Planting in the first week of June will have the highest percentage of pods or seeds in grain legumes. Tighter and wider gap performance in different varieties and sowing after this date have reduced grain yield [9]. Growing green peas at a distance of 20 cm × 10 cm is more sufficient to get good yield benefits [10]. In general, Kharif cultivation requires greater spacing and a smaller plant population than summer crops due to relatively warm temperatures, long vegetative growth, and abundant branching.

3.6 Seed priming

Seed soaking in water for 10–12 hours and shade drying enhances germination percentage and early emergence under rainfed cropping. The seed priming in organic farming refers to soaking the seed in organic liquids. In the broadcast method, the established seed was low so seed soaking and priming a good tool to achieve the desired yield. Gupta and Bhowmick [11] stated that the sowing of pre-sprouted seeds significantly increased the pod number per plant, seed per pod, and test weight in Lathyrus.

3.7 Nutrient management strategies in organic legume production

No doubt, legumes are a very important food crop to improve human nutrition having high protein and nutrient content. Biological response of legumes to plant growth-promoting bacteria (PGPB) or bio-fertilizers is an effective and environmentally safe approach to reduce dependence on chemical and inorganic fertilizers causing soil pollution. Phosphorous (P) deficiency in soil is often a major factor in poor legume productivity. The productivity of legumes can be significantly increased by applying P based on soil analysis information. P applied to legumes can produce residual effects up to a limit of 20 to 35 kg P_2O_5 ha^{-1}. Seeding or seed application with P-soluble organisms such as PSB and PSF increases P-use-efficiency. The level of response to applied P can be further improved by timely management, planting, and source used, maintain optimal plant populations, maintain adequate moisture, low pest incidence, and weed control effectively [12]. Treating seed with bio-fertilizer (rhizobia) can increase legume productivity by 10–12%. The combined use of microbial cultures, such as rhizobium and P based biofertilizer (VAM), leads to higher seed yields of the crop than rhizobia cultures alone. Although good results have been achieved at the research stations, the adaptability of integrated nutrient management (INM) technologies to farmers at the farm level is during early phases of organic farm formations. More attention should be paid to the development and identification of suitable bio-fertilizer strains for major legume systems for different agronomic conditions through integrated methods developed by agronomists, biologists, and microbiologists. The availability of biofertilizer is easy and at cheap rates to the organic farms so better adaptability by farmers to this technology of nutrient management [13]. Several studies have shown that spraying panchgavya at 3% helps improve plant growth because it contains micro and macronutrients and favorable growth hormones. The enzyme in panchagavya promotes rapid cell division and multiplication, helping to improve plant growth patterns. Kumaravelu and Kadambian [14] reported that spraying panchgavya (3%) 10 days after planting (DAS) significantly increased the growth of Greengram plants, resulting in higher grain yield. Several other liquid organic manures such as jeevamrit are also a good option for nutrient management in organic legumes.

3.8 Organic pest and disease management in organic grain legumes

The main issue is the organic management of insects and pests in organic farming. Natural predators should be encouraged and protected (for example, farming trees on farms that attract pests and insects, etc.). Products collected from local farms, animals, plants, and microorganisms and prepared on the farm can control pests. Extracts from neem seed core (NSKE), cow urine sprays are beneficial for pest control. Products authorized to control pests are neem oil and other neem preparations such as NSKE, color traps, mechanical traps, pheromone traps, herbal remedies, mild soaps, and clay, etc. A mixture of 3–5 L cow urine and an equal amount of manure (3–5 kg) stored for 4 days and filtered it. Add 200 g of lime and obtain up to 80 L and spray along with cow's milk in legumes control mosaic, a type of viral diseases (also called yellow mosaic virus) in which whey acts as a good barrier for reducing transmission of YMV [2]. Legumes are susceptible to pests and diseases. Productive losses due to the absence of plant protection measures vary from 46 to 96% depending on the crop and variety in legumes and further in case of organic legumes it can be increased up to complete crop failure. Integrated pest management (IPM) in legumes includes the use of resistant varieties, crop rotation with non-host plants, etc. also a good approach in the management of diseases and pests on organic farms. In Arunachal Pradesh, straw humus reduced the incidence of Ascorobta on peas as mulch application by regulating temperature and humidity [15]. Caterpillars are the most devastating legume insect due to favorable weather conditions. So regular inspections of farms on a daily or weekly basis were implemented.

3.9 Impact on soil dynamics: physical, chemical, and biological properties

Intensive agricultural practices lead to degradation of soil aggregates, resulting in soil erosion and loss of soil organic matter (SOM) [16]. Legume crops increase SOM, help aggregate soil particles, and crop residues obtained from legumes are considered a good technique for sustainable soil management because it prevents soil erosion, improve water holding capacity and help restore soil biodiversity.

3.9.1 Physical properties

Legumes in the form of mulch or crop residue act as a soil conditioner, as they positively affect soil microbial populations by providing them with a substrate, increasing the degradation of the debris or residues of plant and the addition of organic matter to the soil in large quantities [17] lead to the prevention of soil/wind erosion, improved soil agglomeration and water holding capacity, etc. Therefore, legumes play a very important role in soil restoration and provide a favorable strategy to improve soil health for sustainable agriculture.

3.9.2 Chemical properties

Legume cultivation has a tremendously positive effect on soil chemistry. In particular, the pH of the soil decreased due to the production of organic acids and CO_2 from legumes degradation in alkaline soils [18]. Therefore, legumes act as a buffer for the soil by maintaining its pH. The addition of legume crops to intensive farming systems significantly reduces soil and water pollution. Crop residue incorporation based on legumes improves the availability of macro or micronutrients also.

3.9.3 Biological properties

The practice of intercropping with legumes helps in the development of different types of roots capability of fixing nitrogen and is therefore responsible for changing the complete distribution of the roots, as well as the root architecture, as well as modifying the secretion process in the rhizosphere. Therefore, it can strongly influence the microbial community as well as its interactions with the crop, thus promoting various benefits. Intercropping of cereals and legumes also promotes replenishment and facilitation in agricultural systems [19]. Legumes are also thought to promote the invasion of *arbuscular mycorrhizae* (AD) in low-input input agricultural systems, as legumes actively participate in trilateral symbiotic relationships between different microbial species [20]. There is a gradual improvement in the diversity and abundance of mycorrhizae in legume cultivation.

4. Biological N-fixation by grain legumes

The grain legumes can fix nitrogen biologically with the help of a symbiotic and mutual partnership with rhizobia bacteria. N-fixation of grain legumes is about 1.0 kg ha day^{-1} within a cropping season. Most of the excess fixed nitrogen is completely utilized by the second relay crop or it will help in decrease nitrogen demand of the next crop. Legumes can fix up to 100 kg nitrogen per hectare depending upon crop type, management, and agro-climatic conditions. Mostly all grain legumes fix nitrogen such as gram (26–63 kg ha^{-1}), pigeon pea (68–200 kg ha^{-1}), mungbean (50–55 kg ha^{-1}), and lentil (35–100 kg ha^{-1}) [21].

Crop	Wheat	Rice	Lentil
N %	0.39	0.48	1.74
P%	0.13	0.16	0.16
K%	1.10	1.03	1.74

Table 2.
Nutrient concentration and C:N ratio of rice, wheat, and lentil straw.

5. The remaining impact of legumes on agricultural systems

Crop rotation can improve biomass production and can sequester carbon (C) and nitrogen (N) in the soil, mainly crop rotation with non-legumes, and C and N can be improved with effects consecutively, for example, the increase in maize yield in legume rotation is a 50% successful plantation when grown with sesame [22]. Intercropping with legumes may increase soil N, and this total N may not be available during the current growing season, improving soil fertility for successful cultivation [23]. Lopez and Mundt [24] observed that velvet bean (*Mucuna pru-riens*), and sun-hemp (*Crotalaria juncea*) often resulted in maize yields of 4 to 7 mg/ha even without applying additional nitrogen fertilizer in the next harvest. Yusuf et al. [25] reported that to maximize the contribution of legumes N to the next crop, it is essential to maximize the total amount of N in legumes, the amount of N mineralized legumes, and the effectiveness of legumes. The use of this mineral N and the amount of N derived from immobilization Nair et al. [26] observed that legumes, especially soybeans, cowpea, peas, and peanuts intercropping with maize have the remaining or residual effect on the yield of a subsequent wheat crop. Intercropping of sorghum with peanuts, cowpea, and green gram reduces the N demand of wheat fertilizer to 61, 83, and 38 kg/ha for a target yield of 4.0 tons/ha. Grain legume crops have deep root systems in the soil so they can recycle crop nutrients that are deep in the soil profile and also have the ability to hold different nutrients in high amounts in their biological content (**Table 2**).

6. Grain legumes also act as intercropping and green manure in organic farming

Intercropping is practiced to meet different ecological goals, such as promoting interaction between species, activating natural regulatory mechanisms, increasing biodiversity, and reducing farmers' risk against climate aberration [27]. The main objective of the legume intercropping system is to produce more yields from the same field and improve the efficiency of natural resources compared to monoculture [28]. Mixing a legume with a non-legume species may have a performance advantage over monoculture. Legumes improve soil function through the symbiosis of the legume-rhizobia [29]. The main objective of the legume intercropping system is to produce more yields from the same field and improve the efficiency of natural resources compared to monoculture [28]. Mixing a legume with a non-legume species may have a performance advantage over monoculture. Under favorable environmental conditions, legumes add N to the system, obviously leading to high yields of major crops [30]. Barbosae et al. [31] reported that 25% of the fixed N per cowpea component was transferred to corn. An important goal of intercropping is to ensure higher yields per unit area than monoculture [32]. Compost refers to the inclusion of crumbling fresh plant remnants in soils undergoing biodegradation using the soil microbiota and contributing to soil organic matter. Legumes are an effective green manure crop because the decomposed plant matter after harvest can improve soil water retention and water retention, reduce soil erosion and increase SOM, and thus improve soil properties. Different benefits can be used to enhance crop yields [33]. Green manure can be classified into two categories according to their manure position, *i.e.* local green manure, and green manure (Onsite and offsite). In the case of local green manure, legumes are planted and added to the soil on the same site, while in the case of green manure; legumes' waste is collected at the nearest location and added to the soil first, plant the next harvest. Besides, legumes have a wide range of properties such as biological nitrogen fixation, short time, against abiotic and biological stress, environmental flexibility, fast, simple off-farm production, etc. Better monitoring of agricultural sustainability is achieved by legumes addition in organic farming [34].

7. Microbiological changes under organic legumes

Legumes are one of the important components to increase soil microbial biomass (SMB) in the soil [35]. Legumes play an important role in small and medium-sized businesses and important energy processes, such as the nutrient cycle and disruption of SOMs, thus improving crop yields and soil sustainability [36]. The instability of small and medium-sized enterprises implementing several important agronomic processes in the soil can drastically change agricultural productivity and soil sustainability [37]. The relationship between biota and legume in the soil and their significance for the various soil functions has a positive impact on soil sustainability [38]. SMB was increased by rotation of legumes with significant improvements in the structure of soil microbial communities and soil health [39]. Some microorganisms that interact physically with legumes in the rhizosphere can also actively improve crop yields by increasing plant growth and growth [40]. The SMB is similar to the eye of a needle than any SOM must overcome [41] and is therefore widely used as a biological indicator in soil sustainability [38]. SOM is an instant sink for nutrients, organisms, and carbon. SMB also increases the nutrient supply in cultures in symbiotic associations. It contributes to the physical structure of the soil, chemical processes, and pesticide degradation and prevents soil pathogens [42]. SMB and microbiological dynamics are related indicators of changes in soil sustainability due to changes in soil properties. SMBs are mainly found in surface layers and vary according to soil configuration. SMB is a living element of the soil. Tilak [43] stated higher counts of actinomycetes, bacteria, *Azotobacter, fungi,* and PSB due to growing of mungbean in fallow after rice (**Table 3**). These increments in the microbial population in turn affect mineralization and immobilization of nutrients depending upon the environment. By adding legumes in the cropping system, the

Treatments	Microbial population (per g soil)					Soil depth (cm)	
	Bacteria 10^5	Actinomycetes 10^4	Fungi 10^4	*Azotobacter* 10^2	PSB 10^2	0–15	15–30
Rice-fallow	42	0.3	0.1	22	0.4	192.1	156.5
Rice-MB (SR)	105	1.2	0.8	87	3.5	200.5	155.5
Rice-MB (SI)	167	5.5	1.3	202	6.0	244.0	195.7
C D (P = 0.05)	40.5	1.25	0.72	25.8	0.9	35.58	21.24

Table 3.
Soil microbial population as affected by legumes.

microbial population of useful microbes increased up to 3 to 4 times as compared to non-legumes additions. Additional soil biological parameters, improved after long term inclusion of legumes in the cropping sequence/cropping system.

8. Limiting factors for low production

In addition to low productivity, growing population, devastating climate change, complexities of diseases and pests, the socio-economic situation of pulse producers, poor storage facilities, etc. They increase the deficit of legumes that are available in the water.

1. **Area**: The growth of legumes is poor in marginal areas with low resource conditions is one of the main reasons for low pulse productivity. About 87% of the legume growing area in the country belongs to the rain cover system. The average rainfall of the main pulse producing states such as Madhya Pradesh (MP), Uttar Pradesh (UP), Gujarat and Maharastra is about 1000 mm and the variable coefficient of rainfall is 20–25%. Water stress is the most frequently cited reason for the poor harvest. Final drought and heat stress lead to forced maturity with low yields. The stress of drought can only reduce seed yield by 50% in the tropics. A leap in productivity can be achieved by applying life-saving irrigation, especially on Rabi legumes grown with remaining moisture. Two genes have been identified: "efl-1" and "ppd" for early flowering and maturation to escape the stress of drought (ICCV-2 in South India). The actual irrigated area under legumes remains stagnant at 13% of the total area. The availability of suitable soil moisture for plant growth depends on rainfall, water holding capacity, and soil depth in rainy areas. In southern India, soil water holding capacity usually limits grain yield to 50% compared to irrigation capacity. In contrast, in vertisol soil, the ability to retain water is larger, leading to a decrease in the growth rate of up to 5–20%. The increased amount of evaporation in southern India during the Rabi season leads to serious restrictions on the implementation of green beans during drought. Another major problem is soil salinity and alkalinity. High salinity and alkalinity in the semi-arid tropics and the Indo-Gangetic plains of the irrigated areas are of particular interest as most grain legumes are sensitive to salinity and alkalinity.

2. **Pests and diseases**: Among fusarium wilt diseases, associated with the root rot complex, perhaps the most common disease causes significant yield losses in mung beans (**Table 4**). *Fusarium wilt*, sterile mosaic as well as *Phytophthora blight*, yellow mosaic, *Cercospora* spp., and white rust on green and urban peas and rust also cause significant damage. 250 species of insects affect legumes in India. Of these, nearly a dozen cause significant damage to crops. On average, 2–2.4 million tons of legumes worth about 6000 rupees are lost each year due to damage from a pest combination.

3. **Problems with blue bulls**: Favored by blue bulls, the area of legumes is transferred to other crops. Legumes are vulnerable to attack by the blue bull in the Indo-Gangetic Plain. Due to the widespread threat, particularly in Uttar Pradesh, Bihar, Madhya Pradesh, Rajasthan and Chhattisgarh, farmers did not allow the possibility of in pulse production. There is no domestic feasible strategy to effectively combat the threat.

4. **Unfavorable climatic conditions**: Poor soil and agro-climatic conditions not only affect the legumes sowing time but also shorten the time of the growing

Crop	Insect-pest	Yield loss (%)	Disease	Yield loss (%)
Chickpea	*Fusarium* wilt and *Ascochyta* blight	50–100	Pod-borer and cutworm	10–90
Pigeonpea	Sterility mosaic virus	20–70	Pod-borer and leaf	70–80
Cowpea and mungbean	Yellow mosaic virus	10–100	Whitefly	20–25
	Rust, wilt and *Sclerotia blight*	20–70	Pod-borer	—
Field pea	Powdery mildew	10–30	Stem and pod-borer	—
Source: Das [44]; Pande et al. [45] and Satyagopal et al. [46, 47].				

Table 4.
Yield loss in major pulses as caused by weeds, diseases, and insect-pests.

season, but also have to deal with cold damage at the start of the period of the vegetative segment, freezing all biological activity for a long time. The sudden rise in temperature not only causes forced maturity but also causes many biological stresses, diseases, and insect pests [48]. Traditionally, the cultivation of the Rabi legume has been delayed until the last week of November and, under extreme circumstances, until the first half of December, for obvious reasons. However, the optimal time for lentils planting is the first half of October. However, some winter legumes, including lentils, are also grown in pairs in eastern India, making it easy to grow even before the rice is harvested.

5. **Abnormal soil conditions**: In general, legumes prefer neutral soil reactions and are very sensitive to acid, salt, and alkaline conditions and most legumes have phosphorus deficiency. Therefore, P requires significant attention in legume production systems [49]. Indian soil, in particular, the soil in the northwest has a high pH unlike in the east and northeast, characterized by acid soils. Due to these soil conditions, micronutrient deficiency manifests itself in acute scarcity. An acute deficiency of zinc, iron, boron, molybdenum, and secondary nutrients such as sulfur, especially in legumes also reduce the productivity in terms of quality and quantity.

6. **Physiological constraints**: Legumes (plants C-3) have low yield potential and form a group of physically inefficient plants compared to cereals (C-4 plants) such as rice, sorghum, and corn [50].

Crop	Biotic stress	Abiotic stress
Chickpea	Weeds, *Fusarium* wilt, root rot, chickpea stunt, gray mold and pod-borer	Low temperature and nutrient stress
Pigeonpea	Weeds, *Fusarium* wilt, mosaic and pod-borer complex	Cold, terminal drought and waterlogging
Mungbean	Mosaic virus, root and stem rot, stem *Agromyza and* sucking insect-pests stress	Pre-harvest sprouting and temperature
Lentil	*Fusarium* wilt, root rot, and rust	Moisture and temperature
**[52].*		

Table 5.
Important abiotic and biotic stresses limiting the production of major legumes crops in India.

7. **Legume marketing**: The legumes market is fragmented due to the dispersion of production and consumption in all states. Farmers/village traders sell their market surplus immediately after harvesting due to storage losses of pests, while some large traders/wholesalers trade between large markets and stock in legumes to make a profit from speculative profits during the off-season. For this reason, farmers did not take advantage of the higher market prices for pulse [51].

8. **Abiotic stress in terms of crops:** Abiotic stress is mainly inevitable and the most damaging factor for the growth and productivity of legume crops, especially in non-irrigated areas (**Table 5**). The capacity to effectively tolerating this type of stress by the adoption of suitable strategies based on research trails. Fixation, absorption, and assimilation of nitrogen by legumes are reduced due to a reduction in hemoglobin in the nodules and the number of nodules under water stress conditions [53].

9. Recent developments and legume production policies

9.1 Livestock project

Application of biotechnology tools to create a genetic modification for biotic and abiotic stresses and develop varieties suitable for early maturation for late planting situations to escape final water stress and used in some crop systems [54].

9.2 Planting techniques on high beds

Planting on a tall bed is an effective agronomic intervention, especially in areas with heavy rainy season legumes like pigeon peas, chickpeas, and beans. In pigeon peas, the method of planting strips and beds also reduced the incidence of *Phytophthora blight* [55]. Besides, better drainage under growing plantations also reduces the risk of root and stem rot [56]. It helps with controlled irrigation in beds and saves irrigation water (up to 30%) and expensive inputs such as seeds and fertilizers [57]. An increase in the productivity of legumes such as pigeon pea, urdbean, and chickpea was also observed due to the nodding and better growth of culture on a large bed [58].

9.3 Agricultural conservation (CA)

Crops such as summer mung beans can be a sequential partner of crop systems in April–June because legumes have a "preservative" effect on soil nitrogen (N) compared to non-bean crops and are beneficial for crops after crop. Pigeon-wheat and mung-bean systems have shown a clear advantage over wheat-wheat in conservation agriculture. No-tillage and retention of surface residues (tillage for soil conservation) improve the productivity of mung beans compared to conventional tillage practices [57]. Besides, legumes do not need to plow and retain crop residues that provide habitat for beneficial organisms by providing C substrate to heterotrophic microorganisms and increasing microbial activity and improve soil C and N [59].

9.4 Future outlook of organic grain legumes

The literature review presented here identifies that organic legume planting has great potential to promote soil sustainability and organic farming. Organic legumes' ability to improve soil properties (e.g. physical, chemical, and biological) makes them necessary to achieve sustainability goals. Given the immense promise of legumes as a soil amendment, the uncertainties described above must be handled objectively. There is an urgent need to understand the future needs and role of organic legumes in soil sustainability and food security and nutrition.

10. Conclusion

In the modern world, the demand for organic products is increased day by day in such products organic legume had also a great importance. In major organic farms, the adaptability of organic legumes is less as compared to other crops; even they had the capability to acts as a biological nitrogen fixer and act as green manure to other organic crops. There are a lot of abiotic and biotic factors that will adversely affect the productivity and yield of organic legumes. But these stress factors will be compressed by the adoption of suitable techniques on organic farms. This aside, proper market facilities, minimum support prices, and bridgeable gaps reduction and post-harvest management are important for adoption of organic farming in grain legumes and special subsidies in the form of monetary and non-monetary input along with crop insurance can help the farmers to adopt organic legume production without any hesitation.

Author details

Amanpreet Singh* and Harmandeep Singh Chahal
Department of Agriculture, Khalsa College Amritsar, Punjab

*Address all correspondence to: nandgharia@gmail.com

References

[1] Babu S, Singh R, Avasthe RK, Yadav GS. Strategies for enhancing the productivity of organic agriculture in the north eastern region of India. In: Souvenir and Conference Book of National Conference on Global Research Initiatives for Sustainable Agriculture and Allied Sciences, held at Gwalior, Madhya Pradesh, India, 12-13 December 2015. pp. 55-58

[2] Avasthe RK, Singh R, Babu S. Organic legumes production in India: Perspectives and opportunities. Indian Journal of Agronomy. 2016;**61**(Special issue):93-101

[3] APEDA. Annual report of APEDA regarding legymes and pulses. 2019

[4] Pooniya V, Choudhary AK, Dass A, Bana RS, Rana KS, Rana DS, et al. Improved crop management practices for sustainable pulse production: An Indian perspective. Indian Journal of Agricultural Sciences. 2015;**85**(6):747-758

[5] Ali M. Consolidated Report on *Kharif* and *Rabi* Legumes, Agronomy 1997-1998, All India Pulse Improvement Project. Kanpur, Uttar Pradesh: Directorate of Legumes Research; 1998

[6] Yadav A, Suri VK, Kumar A, Choudhary AK, Meena AL. Enhancing plant water relations, quality, and productivity of pea (*Pisum sativum* L.) through AM fungi, inorganic phosphorus, and irrigation regimes in a Himalayan acid Alfisol. Communications in Soil Science and Plant Analysis. 2015;**46**(1):80-93

[7] Sharma DP, Singh MP, Gupta SK, Sharma NL. The response of pigeon pea to short term water stagnation in a moderately sodic soil under field conditions. Journal of the Indian Society of Soil Science. 2005;**53**(2):243-248

[8] Pooniya V, Choudhary AK, Sharma SN, Bana RS, Rana DS, Rana KS. Mungbean (*Vigna radiata*) residue recycling and varietal diversification for enhanced system productivity and profitability in basmati rice (*Oryza sativa*)–wheat (*Triticum aestivum*)–mungbean cropping system. In: Proceedings of National Symposium on Crop Diversification for Sustainable Livelihood and Environmental Security, held during 18-20 November 2014 at PAU, Ludhiana. 2014. pp. 629-630

[9] Padhi AK. Effect of sowing date and planting geometry on a yield of red gram (*Cajanus cajan*) genotypes. Indian Journal of Agronomy. 1995;**40**(1):72-76

[10] Kumar R, Singh M, Waldia RS. Opportunities for extension in area and production of spring/summer mungbean (Vigna radiata) in Haryana (India). Improving income and nutrition by incorporating mungbean in cereal fallows in the indo-Gangetic Plains of South Asia, DFID Mungbean project for 2002-2004. In: Proceedings of Final Workshop and Planning Meeting, Punjab, 27-31 May 2004. Ludhiana, Punjab, India: Agricultural University; 2006. pp. 236-245

[11] Gupta S, Bhowmick MK. Scope of growing Lathyrus and lentil in relay cropping systems after rice in West Bengal, India. Newsletter. 2005;**4**:28-33

[12] Shweta S, Malik M. Response of pulse production to phosphorus—A review. Agricultural Reviews. 2014;**35**(4):295-298

[13] Malhothra RS. Achievements and current status of BNF research in Asia-Pacific. In: International Workshop on Biological Nitrogen Fixation for Increased Crop Productivity, Enhanced Human Health, and Sustained Soil Fertility, 10-14 June. Montpellier, France: INSA-INRA; 2002

[14] Kumaravelu G, Kadambian D. Panchagavya and its effect on the

growth of the green gram cultivar K-851. International Journal of Plant Sciences. 2009;**4**(2):409-414

[15] Choudhary RG. Effect of mulching on Ascochyta blight of pea. In: Annual Report 1990-91. Umiam, Meghalaya: ICAR Research Complex for NEH Region; 1990. p. 43

[16] Meena RS, Yadav RS, Reager ML, De N, Meena VS, Verma JP, et al. Temperature use efficiency and yield of groundnut varieties in response to sowing dates and fertility levels in the Western dry zone of India. American Journal of Experimental Agriculture. 2015;**7**(3):170-177

[17] Lal R. Sequestering carbon in soils of agro-ecosystems. Food Policy. 2011;**36**:S33-S39

[18] Mitran T, Meena RS, Lal R, Layek J, Kumar S, Datta R. Role of soil phosphorus on legume production. In: Meena et al., editors. Legumes for Soil Health and Sustainable Management. India: Springer; 2018. DOI: 10.1007/978-981-13-0253-4_15

[19] Duchene O, Vian JF, Celette F. Intercropping with legume for agroecological cropping systems: Complementarity and facilitation processes and the importance of soil microorganisms; a review. Agriculture, Ecosystems and Environment. 2017;**240**:148-161

[20] Pivato B, Mazurier S, Lemanceau P, Siblot S, Berta G, Mougel C, et al. *Medicago* species affect the community composition of arbuscular mycorrhizal fungi associated with roots. The New Phytologist. 2007;**176**:197-210. DOI: 10.1111/j.1469-8137.2007.02151.x

[21] Gills MS, Prasad K, Ahalawat IPS. Improving the sustainability of rice-wheat cropping system through pulses: Weeds and imperatives. In: Ali M, Gupta S, Basu PS, Naimuddin, editors. Legumes for Ecological Sustainability. Kanpur: Indian Society of Pulses Research and Development; 2009. pp. 71-91

[22] Kureh I, Kamara AY. Effects of sole cropping, intercropping, and rotation with legume trap-crops on Striga control and maize grain yield in farmers' fields in the northern Guinea Savanna. In: Badu-Apraku B, MAB F, Lum AF, Menkir A, Ouedraogo M, editors. Demand-driven technologies for sustainable maize production in West and Central Africa. Fifth Biennial West and Central Africa regional maize workshop, 3-6 May 2005, IITABénin. 2005. pp. 169-179

[23] Lithourgidis AS, Dordas CA, Damalas CA, Vlachostergios DN. Annual intercrops: An alternative pathway for sustainable agriculture. Australian Journal of Crop Science. 2011;**5**:396-410

[24] Lopez CG, Mundt CC. Using mixing ability analysis from two way cultivar mixtures to predict the performance of cultivars in complex mixtures. Field Crops Research. 2000;**68**(2):121-132

[25] Yusuf AA, Iwuafor ENO, Olufajo OO, Abaidoo RC, Sanginga N. Effect of crop rotation and nitrogen fertilization on yield and nitrogen efficiency in maize in the northern Guinea Savanna of Nigeria. African Journal of Agricultural Research. 2009;**4**(10):913-921

[26] Nair KPP, Patel UK, Singh RP, Kaushik MK. Evaluation of legume intercropping in the conservation of fertilizer nitrogen in maize culture. Journal of Agricultural Science. 1979;**93**(1):189-194

[27] Kumar S, Sheoran S, Kumar SK, Kumar P, Meena RS. Drought: A challenge for Indian farmers in context to climate change and variability. International Journal of Scientific Progress & Research. 2016;**11**:6243-6246

[28] Inal A, Gunes A, Zhang F, Cacmak I. Peanut/maize inter-cropping induced changes in rhizosphere and nutrient concentrations in shoots. Plant Physiology and Biochemistry. 2007;**45**:350-356

[29] Fustec J, Lesuffleur F, Mahieu S, Cliquet JB. Nitrogen rhizodeposition of legumes: A review. Agronomy for Sustainable Development. 2010;**30**:57-66

[30] Ghaley BB, Hauggaard-Nielsen H, Hogh-Jensen H, Jensen ES. Intercropping of wheat and pea as influenced by nitrogen fertilization. Nutrient Cycling in Agroecosystems. 2005;**73**:201-212

[31] Barbosae S, Lima PI, Silva PS, de Oliveira OF, de Sousa RP. Planting times of cowpea intercropped with corn in the weed control. Revista Caatinga. 2008;**21**(1):113-119

[32] Raji JA. Intercropping soybean and maize in a derived savanna ecology. African Journal of Biotechnology. 2007;**6**(16):1885-1887

[33] Dinnes DL, Karlen DL, Jaynes DB, Kaspar TC, Hatfield JL, Colvin TS, et al. Nitrogen management strategies to reduce nitrate leaching in tile-drained Midwestern soils. Agronomy Journal. 2002;**94**:153-171

[34] Sofi PA, Baba ZA, Hamid B, Meena RS. Harnessing soil Rhizobacteria for improving drought resilience in legumes. In: Meena RS et al., editors. Legumes for Soil Health and Sustainable Management. Springer; 2018. DOI: 10.1007/978-981-13-0253-4_8

[35] Siddique KHM, Johansen C, Kumar Rao JVDK, Ali M. Legumes in sustainable cropping systems. In: Kharkwal MC, editor. Food Legumes for Nutritional Security and Sustainable Agriculture. Proceedings of the fourth international food legumes research conference (IFLRC-IV), October 18-22, 2005, New Delhi, India. Vol. 1. New Delhi: Indian Society of Genetics and Plant Breeding; 2008. pp. 787-819

[36] Knight TR, Dick RP. Differentiating microbial and stabilized β-glucosidase activity relative to soil quality. Soil Biology and Biochemistry. 2004;**36**:2089-2096

[37] Devare M, Londono RLM, Thies JE. Neither transgenic Bt maize (MON863) nor tefluthrin insecticide adversely affect soil microbial activity or biomass: A 3-year field analysis. Soil Biology and Biochemistry. 2007;**39**:2038-2047

[38] Leterme P, Carmenza Munoz L. Factors influencing legume consumption in Latin America. The British Journal of Nutrition. 2002;**88**(Suppl 3):251-254

[39] Chintala R, Mollinedo J, Schumacher TE, Malo DD, Julson JL. Effect of biochars on chemical properties of acidic soil. Archives of Agronomy and Soil Science. 2013;**60**:393-404

[40] Deakin WJ, Broughton WJ. Symbiotic use of pathogenic strategies: Rhizobial protein secretion systems. Applied Soil Ecology. 2009;**7**:312-320

[41] Prell J, Poole P. Metabolic changes of rhizobia in legume nodules. Trends in Biotechnology. 2006;**14**:161-168

[42] Liang B, Lehmann J, Sohi SP, Thies JE, O'Neill B, Trujillo L, et al. Black carbon affects the cycling of non-black carbon in the soil. Organic Geochemistry. 2010;**41**:206-213

[43] Tilak KVBR. Response of sesbania green manuring and mungbean residue incorporation on microbial activities for sustainability of a rice-wheat cropping system. Journal of Agriculture and Rural Development in the Tropics and Subtropics. 2004;**105**(2):189-196

[44] Das TK. Weed Science: Basics and Applications. Jain Brothers, New Delhi: India; 2008. pp. 666-679

[45] Pande S, Sharma M, Kumari S, Gaur PM, Chen W, et al. Integrated foliar diseases management of legumes. In: Proceedings of International Conference on Grain Legumes: Quality Improvement, Value Addition, and Trade, 14-16 February 2009. Kanpur, Uttar Pradesh, India: Indian Institute of Pulses Research; 2009. pp. 143-167

[46] Satyagopal K, Sushil SN, Jeyakumar P, Shankar G, Sharma OP, Boina, et al. AESA based IPM Package for Redgram. Rajendranagar, Hyderabad, India: National Institute of Plant Health Management; 2014a

[47] Satyagopal K, Sushil SN, Jeyakumar P, Shankar G, Sharma OP, et al. AESA Based IPM Package for Blackgram and Greengram. Rajendranagar, Hyderabad, India: National Institute of Plant Health Management; 2014b. p. 43

[48] Ali M, Gupta S. Carrying capacity of Indian agriculture: Pulse crops. Current Science. 2012;**102**(6):874-881

[49] Kumar A, Suri VK, Choudhary AK. Influence of inorganic phosphorus, VAM fungi, and irrigation regimes on crop productivity and phosphorus transformations in okra (*Abelmoschus esculentus L.)–pea (Pisum sativum L) cropping* system in an acid Alfisol. Communications in Soil Science and Plant Analysis. 2014;**45**(7):953-967

[50] Reddy AA. Legumes production technology: Status and way found. Review of Agriculture. 2009;**XLV1**(52):451

[51] Naryan, Kumar. Constraints of growth in area production and productivity of legumes in India: An analytical approach to major legumes. Indian Journal of Agricultural Research. 2015;**49**(2):114-124

[52] Reddy AA. Impact Assessment of Pulses Production Technology, Research Report No 3. Kanpur, Uttar Pradesh, India: Indian Institute of Pulses Research; 2006

[53] Krishnamurthy L, Kashiwagi J, Gaur PM, Upadhyaya HD, Vadez V. Sources of tolerance to terminal drought in the chickpea (*Cicer arietinum* L.) minicore germplasm. Field Crops Research. 2010;**119**(2-3):322-330

[54] Umesha C, Sridhara CJ, Kumarnaik AH, Shivarajkumar HS. Ways to bridge yield gaps and production problems in pigeonpea cropping systems. Journal of Pharmacognosy and Phytochemistry. 2017;**6**(5):2651-2657

[55] Chauhan VB, Singh VB, Singh ON. Integrated management on *Phytophthora* blight of pigeon pea. Legume Research - An International Journal. 2005;**18**(1):46-49

[56] Chaudhary RG, Naimuddin. Resource conservation technologies and integrated disease management strategies in legumes. In: Ghosh PK, Kumar N, Venkatesh MS, Hazra KK, Nadarajan N, editors. Resource Conservation Technology in Legumes. Jodhpur: Scientific Publisher; 2014. pp. 535-553

[57] Kumar N, Singh MK, Ghosh PK, Hazra KK, Venkatesh MS, Nadarajan N. Resource conservation technology in pulse-based cropping systems. In: Technical Bulletin. Kanpur: Indian Institute of Legumes Research; 2012. pp. 1-35

[58] Kumar N, Singh MK, Praharaj CS, Singh U, Singh SS. Performance of chickpea under different planting methods, seed rate, and irrigation level in indo-Gangetic Plains of

India. Journal of Food Legumes. 2015;**28**(1):40-44

[59] Bhattacharyya R, Das TK, Pramanik P, Ganeshan V, Saad AA, Sharma AR. Impacts of conservation agriculture on soil aggregation and aggregate-associated N under an irrigated agro-ecosystem of the Indo-Gangetic plains. Nutrient Cycling in Agroecosystems. 2013;**96**:185-202

Chapter 6

Sustainable Production of Pulses under Saline Lands in India

Anita Mann, Ashwani Kumar, Satish Kumar Sanwal and Parbodh Chander Sharma

Abstract

The decreasing agricultural lands along with waste lands and poor water resources are the main constraints for sustainable agricultural production. The need of time is to produce maximum with minimum inputs. Depleting levels of major and micro-nutrients in Indian soils have been on the rise, and situation may be more harmful if corrective measures are not followed in time. The soil nutrient deficiencies significantly reduce the crop yields in addition to the soil fertility. In preview of this, the need of the hour is to conserve agricultural sustainability, soil health enhancement, and water management. Farmers are forced to use saline water for irrigation in areas with poor quality water or less available water for irrigation, specifically in arid or semi-arid regions. Every crop plants have threshold limit of tolerance beyond which salinity decreases the crop yield. Legumes are very sensitive crops towards soil salinity, and secondary salinization mainly through irrigation water is the hardest challenge for survival of legume crops in arid regions. In view of this, the sustainability of legumes in salt affected areas is a big challenge for crop productivity being sessile to salinity. Hence, the possible strategies for sustainability of salt sensitive legumes have been briefly reviewed in this chapter.

Keywords: soil salinity, pulses, abiotic stress, tolerance, sustainability

1. Introduction

Pulses are one of the food crops that address the challenge of global food security and climate changes, as well as contribute to healthy diets. In recognition of the importance of pulses for human well-being, year 2016 was declared as the International Year of Pulses by the UN General Assembly. Pulses are a subgroup of legumes belonging to Leguminosae family and its seeds are edible FAO [1]. All legumes are sometimes not categorized as pulses e.g., oil producing soybean (*Glycine max* [L.] Merr.) does not come under pulses but grouped as oil-seed crop. Similarly, some other legumes which are consumed as green vegetables e.g., *Medicago sativa* L. are also excluded from pulses. Pulses can be defined as protein rich plants that contribute to healthy diet for people across the globe. Nutritionally, pulses are rich source of dietary fiber, minerals and vitamins with low levels of cholesterol and fat. The protein content in pulses is more than double in comparison to cereals and hence are the important source for nutritional and balanced diet for humans. Pulses are included in all "food baskets" and dietary guidelines. The World Food Program (WFP), for instance, includes 60 g of pulses in its typical food basket, alongside cereals, oils and sugar and salt. India is the largest producer of pulses in the world, with 25% share in the global production. In India, mainly mung bean or green gram (*Vigna radiata*), chickpea or Bengal gram (*Cicer arietinum*), urad bean or black gram (*Vigna mungo*), moth bean (*Vigna aconitifolia*), pigeonpea or red gram (*Cajanus cajan*), lablab bean (*Lablab purpureus*), broad bean or faba bean (*Vicia faba*), horse gram (*Dolichos uniflorus*), lentil (*Lens culinaris*), pea (*Pisum sativum var. arvense*), grass pea or khesari (*Lathyrus sativus*), cowpea (*Vigna unguiculata*) etc. are grown as pulses. Generally, two growing seasons are available for pulses kharif and rabi. Rabi season legumes chickpea, lentil, and dry peas are the cool, dry season pulses grown during October–April while pigeonpea, urad bean, mung bean, and cowpea are grown preferably during the warmer, rainy season or kharif from June to October [2, 3]. The global legume/pulse production, area and yield during 2013 was ~73 million tons (MT), ~80.8 million ha (M ha) and ~904 kg ha-1respectively [4]. Further, Africa and Asia together contribute ~49 MT, i.e., 67% of the global pulse production. In India, production of pulses is around 25.23 million tons during 2017–2018 (DAC, 2018) with maximum production of soybean (11.95 MT) followed by chickpea (~11.23 MT), groundnut with shell (9.4 MT), pigeonpea (~4.25 MT), lentils (~1.61 MT) and dry peas (0.6 MT). During the last 50 years from 1961 to 2011, overall, an average of ~1.7 times legume production has increased globally with ~3 times more cereal production in the same period. Increase in yield per hectare of legumes was observed only ~1.4 times with ~3.0 times increase in cereals. These data show that there is an acute shortage of pulses in India to meet the daily recommended diet. The area, production, growing states with distribution of pluses in India is being given in **Figure 1**.

The Legume family consists of important food grain, oilseed, forage, and agro forestry species. The domestication of legumes by humans dates back to Neolithic times. Chickpea is one of the seven Neolithic founder crops of the near East [5].

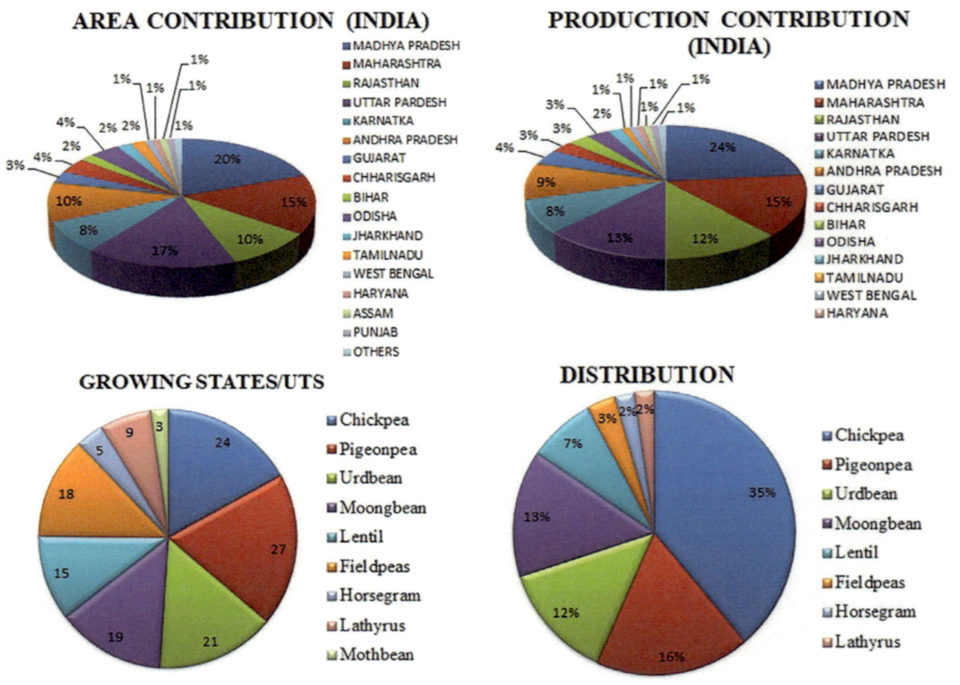

Figure 1.
Domestic production of domestic pulses in different states of India with percent distribution of each legume. Source [11].

Some of the earliest domesticated legumes include: lentil ~9000 years [6], beans and soybean (~3000 year) [7, 8]. Legumes include ~750 genera and ~18,000 species [9, 10]. Legumes form an important part of human daily diet especially in several developing and some developed countries and therefore sometimes legumes are considered as poor man's meat. Pulses contain various amino acids and also have medicinal properties and hence, are consumed as Dal, the easily available source of protein. In addition to the nutrient rich seed/grain, plant parts like leaves, pod coats etc. are used as animal fodder e.g., moong bean, lobia, urad bean and gram. These plants are also used as green organic manure to maintain soil health and fertility.

Legumes are under explored crops in comparison to staple cereal crops like wheat, rice and maize and hence are prioritized over pulses in most of the crop improvement programs. In addition to that pulses being sensitive to biotic and abiotic stresses, farmers have reduced their cultivating due to low yield potentials. Hence, keeping these facts in view, it is required to pay more attention for sustainable production of pulses in arid and semi-arid regions.

India will face intense pressure on its land and water resources in agriculture because of diversion of resources to domestic, industrial and other sectors of economy and the likely degradation of these resources, having to feed 1.6 billion people by 2050. Ever increasing demand for good quality land and water resources in the domestic and industrial sectors has already generated enhanced interest in the utilization of salt affected soils. While salt affected soils currently constitute 6.74 M ha in different agro-ecological regions, the area is likely to increase to 16.2 M ha by 2050. Thus, salt affected soils represent an opportunity that can be exploited to increase agricultural production and productivity to ensure national food and nutritional security. The distribution of salt affected soils occurs mostly in arid and semi-arid regions although such soils may exist in every climatic region including a good area of irrigated lands. In addition, the coastal salinity is another big challenge. Hence, the diversity of soil properties in different agro-climatic conditions requires different approaches to reclaim and maintain the soil properties.

The dynamics in the global pulses sector are ever changing and keeping pace with them is much like running on a treadmill. For the past few decades, India has been trying its best to make pulses more and more accessible to its citizens given the rapid drop over the previous few decades. On the other side, there is increased interest in vegetarianism, healthier diets, apart from falling real incomes following slowing economies and the relative lower cost of protein through pulses, all of which increase demand in countries like US and China. There is an increased "return to traditional foods" in African countries like Egypt and Morocco that are adding to the demand mainly import demand. From 2005 when Spain was the second largest importer, followed by Italy and Mexico, in 2012, the profile of importing countries had moved to China, Pakistan, Bangladesh, Egypt and then Spain. The cheaper access to proteins that pulses provide is clearly important in this age of tough economies.

Pulses import basket now has mainly dry peas resembling palm oil in edible oils. Yet the record domestic chana production and 40% correction in chana price have not been able to compete with peas. Similar is with Pigeon Peas. India traditionally imported dry peas mainly from Canada, but now cheaper imports from Russia, Ukraine and France are also keeping the prices under pressure making it difficult to keep Chana and Pigeon peas competitive. None of the steps taken in the recent years seem to have paid off in terms of better farmer remuneration, or better availability and better prices for consumers. The only real solution to this problem is increasing domestic production to the extent that over 90–95% of the demand is met internally and the imports are supplemental to ensure prices are not volatile. It requires policies that encourage better technologies, better quality, better post-harvest management and better distribution of the pulses.

In the current season chana, moong, tur and urad have all seen production increase with increases in acreages and excellent weather. This may be a good opportunity to also open exports and see if farmer incomes and long term acreages increase with these measures as against the known vicious cycle of acreages decreasing after a season of good production following the crash in prices led precisely by the high production. It may take a few years for the prices to settle and farmers to become globally integrated, but the time has indeed come to take this leap of faith [12].

2. Mode of formation of saline soils

A big monetary loss created due to soil salinity in India. About 175 M ha (53%) land is degraded by various means from a total of 329 M ha of land in the country. Ministry of Agriculture, GOI stated that 7.61 M ha of land is salt affected in India which ranges from 8.56 M ha to 10.9 M ha. Majorily saline—alkali and water-logging problems occur in Gujarat, Andhra Pradesh, parts of Punjab, Haryana, Rajasthan, Maharashtra, Karnataka, Uttar Pradesh, Madhya Pradesh, and Tamil Nadu (**Table 1**). Total salt affected soils in India are 1,710,673 ha with maximum in Gujarat followed by Rajasthan (**Figure 2**) while Coastal saline soils are 1,246,136 ha with maximum in Gujarat followed by West Bengal and Orissa.

Thus briefly following reasons can be assigned for salinization of the soils:

- Non-systematic irrigation with more than required irrigation cycles
- Evapotranspiration in arid conditions leads to accumulation of salts in the top layer

Sr. no.	State	Saline soil (ha)	Alkali soil (ha)	Coastal saline soil (ha)	Total (ha)
1	Andhra Pradesh	0	196,609	77,598	274,207
2	A & N islands	0	0	77,000	77,000
3	Bihar	47,301	105,852	0	153,153
4	Gujarat	1,218,255	541,430	462,315	2,222,000
5	Haryana	49,157	183,399	0	232,556
6	J & K	0	17,500	0	17,500
7	Karnataka	1307	148,136	586	150,029
8	Kerala	0	0	20,000	20,000
9	Maharashtra	177,093	422,670	6996	606,759
10	Madhya Pradesh	0	139,720	0	139,720
11	Orissa	0	0	147,138	147,138
12	Punjab	0	151,717	0	151,717
13	Rajasthan	19,557	179,371	0	374,942
14	Tamil Nadu	0	354,784	13,231	368,015
15	Uttar Pradesh	21,989	1,346,971	0	1,368,960
16	West Bengal	0	0	441,272	441,227
	Total	1,710,673	3,788,159	1,246,136	6,744,968

Table 1.
Extent and distribution of salt affected soils in India.

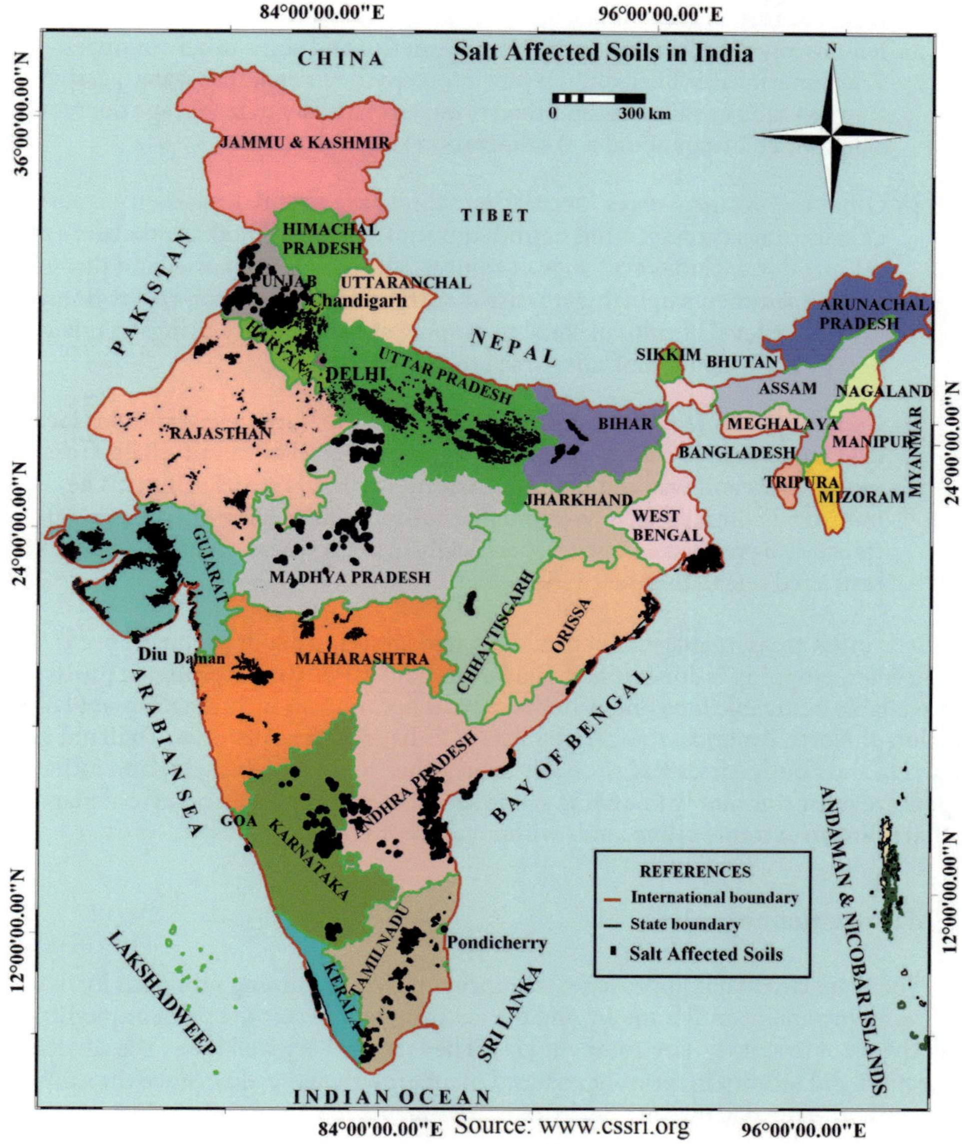

Figure 2.
Distribution of salt affected soils in India.

- Seepage in perennial river basins/irrigation sources causes water logging conditions
- Heavy use of chemical fertilizers containing chlorides, sulfates etc.
- Poor drainage conditions.

i. Use of saline groundwater: the areas where no other source of water is available, irrigation with highly saline groundwater leads to accumulation of salts in plant root surrounding environment. This problem is more prone in soils with poor drainage.

ii. Saline seeps is a major problem in areas with change in land use pattern, when natural forests are being used for cropping system or following a fallow season in green lands or some change in farming system. This problem mainly found in Australia, North America and some other countries. The impermeable horizontal layers intercept the percolating water passing through saline sediments and then transport laterally to landscape depressions which brings about soil salinization [13].

iii. Other important sources for creating salinity are sea tidal waves, transport of salt sprays through wind or underground aquifers. Another procedure of salinization includes exchange of soluble salts between sea and land through marine sediments uplifting on earth's surface. In arid or semi-arid regions, the lowest level of rainfall equal to evapotranspiration also brings salinization by accumulation of salts in the root zone of plants.

iv. Generally soluble salts move from higher levels to lower levels and this localized movement of salts also builds a significant level of salinity. Salts from moist to dry and watered fields to adjacent dry fields create salinity. The industrial/commercial developmental activities like laying of roads or rail tracks also create salinization specifically in areas where natural drainage is restricted due to these activities.

In rainfed areas, management of unirrigated lands is a major problem specifically when cropping is done in such fields. Dryland salinity is a worldwide threat to available cultivable land and water resources in countries including. Great Plains region of North America, Iran, Afghanistan, India, western Australia, Thailand and Canada and South Africa and probably some other countries. Such dryland saline spots are most commonly known as saline seeps, occurring frequently from marginal saline to extreme saline lands without any cultivation.

3. Management of saline soils

There are two major approaches to improving and sustaining productivity in a saline environment-either modifying the environment to suit the plant or modifying the plant to suit the environment [14]. The farmers very well know the adverse effects of soil salinity in terms of reduced plant growth and yields. Since the saline area is increasing day by day, the farmers are shifting to various other alternates like change of cropping pattern, green manuring, dairying etc. along with use of open drainages and mulching. Although it is difficult to manage saline soils through any chemical amendment, the only precautionary method is to use good quality water and grow salt tolerant varieties of crops.

The consideration of various studies on soil salinity have led to only a single observation for timely implementation of corrective measures to stop further salinization and conversion of fertile soils to waste lands. This approach cannot be achieved by a single adaptation, it needs to be a cumulative effort to spread the awareness about soil sustainability and enhanced crop production. In the coming time the concept of soil management and sustainability should be the one of the important issue to be taken care of or we have to bear the low crop productivity (www.nabard.org).

4. Crops and saline soils

Most crops do not grow well on soils that contain salts. Homeostasis of ions is disturbed in the cell due to high salt concentration in the rhizosphere, thereby, creating water deficit under salt stress conditions. This leads to abbreviations in structure and function of various proteins. Under such circumstances, a signaling pathway is activated in the cell to synthesize the metabolites, proteins or enzymes involved directly in scavenging free radicals and maintaining the ionic flux through osmoregulation. Detoxification of free radicals is an important defense mechanism under salt stress. Pulses in general are the most salt sensitive crops. One reason is that salt causes a reduction in the rate and amount of water that the plant roots can take up from the soil. In legumes, salt stress imposes a significant limitation of productivity related to the adverse effects on the growth of the host plant, the root-nodule bacteria, symbiotic development and finally the nitrogen fixation capacity. Also, some salts are toxic to plants when present in high concentration. Some legumes plants are more tolerant to a high salt concentration than others e.g., lentils are more tolerant than soybean and chickpea. Recently, few salt tolerant chickpea lines have been identified for survival at early seedling stage in saline soils with EC of irrigation water from 3 to 12 dS/m [15]. Earlier, pea genotypes have been categorized as sensitive, moderately tolerant or tolerant to salinity based on physiological and biochemical expressions of salt tolerance [16].

5. Pulses and sustainability

The most potential technologies in pulse production include improved crop establishment and management practices, integrated soil fertility and pest management practices, etc., which enhance not only the productivity and profitability but also warrant environmental and social sustainability besides nutritional security [17]. Role of pulses in maintaining sustainability is very wide since pulses are an important component of crop rotations, requiring very less fertilizers than other crops. Legumes are beneficial part of the rotational crops to maintain soil fertility.

Good soil management practices for crop rotation also includes pulses with different crops like wheat, barley, oats, oilseeds such as canola, flax, sunflowers, etc. The nitrogen fixing ability of pulses enriches the soil quality and fertility and hence leading to enhanced productivity in subsequent crop rotations. Soil health in also improved through pulses as they feed soil microbes. Even the crop residue of pulses contains different amino acids and bio-chemicals than non-legumes. This diversity in soil composition provides better protection against disease-causing microbes and thus helps crops to thrive under adverse conditions. The number and diversity of soil microbes is markedly increased with intercropping of pulses [17, 18]. An environment of 'live' soil with diversity of soil micro-organisms is considered best for crops because these micro-organisms enhance the nutrient uptake rate and efficiency in different soils. Additionally, the abundance of diverse soil micro-organisms 'crowd out' the disease-causing bacteria and fungi, and thus protects the plants.

Pulses are considered as low carbon footprint crops since they use half energy inputs than other crops. Soil bacteria are utilized for nitrogen fixation from air which replaces additional requirement of nitrogen fertilizers in pulse crops. Nitrogen enriches the soil in different forms like fertilizer, manure or crop residue, then most part of it is converted into a powerful greenhouse gas, nitrous oxide. Globally, nitrous oxide represents around 46% of the greenhouse gas emissions from agriculture and is almost 300 times more potent than carbon dioxide (CO_2) [19]. As nitrogen fertilizers are related directly with greenhouse gas emissions therefore, pulses have lower carbon footprint than other crops due to their nitrogen-fixing ability. Nitrogen is manufactured from natural gas and is the most needed fertilizer in crop production. The unique feature of pulses is that they take nitrogen from the air directly and fix at their own, hence, application of nitrogen as fertilizer in pulses is minimal in comparison to other crops.

The nitrogen fixation property of pulses reduces the footprint of other crops in the soil adding to the food production cycle. In a study on durum wheat, it was found that carbon footprint was reduced by 17% in durum wheat when preceded by chickpeas or lentils, nitrogen-fixing crops, than preceded by a cereal crops. The impact of pulse-pulse intercropping with wheat system was much stronger than traditional cereal-cereal-durum rotation by reducing the carbon footprint of durum wheat up to 34%.

The water requirement of pulses is also very less i.e., 1/2 to 1/10 of water in comparison to other protein rich crops [20]. Few pulses are well adapted to dry climates and hence can be produced better under dry/drought conditions, e.g., peas and lentils roots absorb less water from a lower depth. The water available in deep soil is, in this way, used by next crop, thereby increasing the water use efficiency of the entire cropping system.

6. Role of legumes in improving soil quality

Legumes have long been recognized and valued as "soil building" crops. Growing legumes improves soil quality through their beneficial effects on soil biological, chemical and physical conditions. When properly managed, legumes will:

- Enrich the soil N power with increased reserved organic matter
- Enhance soil biological activity with improved soil structure
- Increase soil aeration with reduced soil erosion
- Improve water-holding capacity of soil

These improvement factors depend on legume type, remaining plant residue left over in soil along with environmental conditions.

The legumes crops that are capable of fixing large amounts of nitrogen under good moisture conditions, can significantly improve the nitrogen supply for succeeding crops when grown annually. Annual grain legumes are normally grown for grain production along with green manure crops. The selection of legume type and variety mainly depends on minimum support price of the crop, including the climatic adaptability to that particular area (e.g., few varieties of chickpea and lentil are more drought-tolerant than faba bean and pea), adaphic factors, disease tolerance, etc. The effect of perennial forage legumes on soil quality is generally to greater extent and for longer duration than of annual pulse crops. The root structure and growth habit generally determines the amounts of nitrogen fixed by grain legumes and the shallow roots of pulses with short growth cycle become a limitation for their influence on soil physical conditions. However, during the growth of grain legumes, considerable amounts of nitrogen are leaked from roots into the soil. Also, the residues from these crops have a higher nitrogen content than cereal straw and they break down more readily, releasing nitrogen into the soil. Thus, cereal crops that follow grain legumes require less N fertilizer.

Biennial forage legumes e.g., sweet clover, yellow-flowered type is preferred by farmers. It is more drought-tolerant, shorter, and finer stemmed and leaved. These characteristics make it a more palatable livestock feed and easier to incorporate as green manure. Sweet clover is one of the most suitable crops for use as a green

DOI: http://dx.doi.org/10.5772/intechopen.91870

manure. This helps in recovery of soil moisture reserves and residue decomposition during the partial fallow period. Sweet clover is very well suitable on moderately saline soils or soils that have poor structure due to very low levels of organic matter. An experiment on wheat with sweet clover green manure in rotation for 3 years at Indian thin black soils yielded significantly more than with crop rotations with fertilized crops or fellow lands [21, 22]. Although unfertilized green manure-wheat-wheat rotation also yielded highest during the initial years of this study.

Perennial forage legumes also add humus and soil nutrients in addition to fixing nitrogen. Again, the selection of perennial legume depends mostly on the usefulness of crop and soil type.

Alfalfa is a well-known legume tolerant to drought, moderate salinity, winter hardy along with high yields which is widely grown over diverse range of adaphic conditions. Neutral to slightly alkaline soils favor the production and persistence of alfalfa but acidic soils (pH less than 6.0) limit the production severely. The drought tolerance of alfalfa is due to deep root system and hence it grows best on well-drained soils. It cannot grow well on soils with poor drainage and hence is flooding intolerant.

Red clover, a perennial legume, has growth cycle shorter than alfalfa and therefore included in short duration intercropping system. Since, it is well adapted to diverse soil types, it is grown mostly for seed and feed in the moist areas. Red clover is sensitive towards salinity or drought but tolerance level to acidic soils is more than alfalfa.

Forage legumes have deep root system and longer growth period thus have greater capacity for nitrogen fixation and hence improve soil quality more effectively. Biennial and perennial forage legumes add more nitrogen and organic matter in second year after under seeding in cereals in wet areas. Regular green manuring with forage legumes on degraded soils with typically low organic matter contents, adds more nitrogen and organic matter to the soil. It was found that yield of 12 successive wheat crops increased significantly than a non-legume system on a Gray soil in Northern Alberta. The reason may be addition of nutrients and nitrogen to soil from the deep-rooted legume improved soil quality. The benefit of using forage legumes as green manure enriched soil organic matter with nitrogen and readily decomposable plant residue. This also enhances the production of soil microbes and thus, fastens the nutrients availability. It was found [23] that deep-rooted perennial legumes take up phosphorus from the subsoil although available phosphorus is mostly found in the 0–6 inch depth and not below than 2-foot depth. Thus, green manuring with these legumes increases availability of soil nutrients.

In an experiment at Saskatchewan, it was observed that tillage was easier with lower power requirements on soils following a perennial legume crop than after cereal grains. The reason may be improved soil structure with more porosity due to readily decomposable soil organic matter leftover by legumes. As a result, the water holding capacity of soil is increased and it becomes more friable and less erosive and easier to till.

Forage legumes also have the tendency to reduce salinity problems. Alfalfa uses excessive water through its deep roots and thus deep penetration of roots improves the internal soil drainage. A study at Outlook with irrigated alfalfa plowed-down in late fall or early spring indicated that the following cereal crop required little nitrogen fertilizer, while the second cereal required two-thirds of its usual amount (Henry, University of Saskatchewan, Saskatoon).

In view of above facts, key insights and findings from the literature and case studies across the three pillars of sustainability, environmental, social and economic impacts of pulses are summarized below.

7. Environmental impacts

Nitrogen fixation: the unique role of pulses in the global nitrogen cycle is due to their property of fixing atmospheric nitrogen in soils. The intercropping of pulses with traditional crops reduces fertilizer requirements of soil in subsequent crop cycles by actively fixing nitrogen in the soil. The leftover nitrogen in the soil also increases the productivity in subsequent crops. Hence, the benefit of pulses observed in different cropping systems to enhance the soil properties and crop productivity is a matter of an ever-evolving science.

The changing patterns in traditional planting methods of maize and bean affect the nitrogen balance in cropping systems in Sub-Saharan Africa. It was found that, intercropping of cowpea with maize and groundnut enhanced the nitrogen benefits. Also, modified strip-cropping of cowpea and sorghum altered to traditional farming practices prevented the nutrient losses in the dry savannahs of Nigeria and Niger. The environmental benefits of pulses are prominent in terms of their nitrogen fixation abilities, reduction in fertilizer nitrogen requirements and nutritional enrichment (protein content) of succeeding cereal crop strong. Related benefits of the reduced synthetic nitrogen fertilizers requirements in cropping systems, when pulses are added in rotations, include the reduced emissions and energy use associated with the production, use and disposal of fertilizers.

In the twentieth century, the negative effects of human impacts on the global nitrogen cycle are mainly due to increased and imbalanced fertilizer use and fossil fuel combustion causing severe water pollution and high emissions of N_2O. It has been observed that nitrogen fertilizer use has increased by roughly 800% from 1960 to 2000, which is almost half of that being used for wheat, rice, and maize production [24]. Synthetic fertilizers provided close to half of all the nutrients received by crops globally during the mid-1990s, demonstrating both a large dependency on synthetic fertilizers, but also inefficient management of nitrogen in global agriculture [25]. Cereal crops such as wheat, rice and maize typically only utilize 40% of fertilizer applied, leading to significant waste and environmental impacts such as eutrophication of coastal waters and creation of hypoxic zones [24].

It was observed that field pea, lupin or faba bean fulfilled about 70% of nitrogen requirements from atmospheric nitrogen with an average of 19 kg of nitrogen fixed per ton of pulse shoot dry matter during 2001–2013. The study was conducted across the geographic range of southern and central New South Wales, Mallee and Wimmera in Victoria, and the high rainfall zone of south-eastern South Australia [26]. The two most important interventions to decrease nitrogen requirement are systematic crop rotation including pulses/legumes into maize-based systems and optimizing the timing of application with balanced amounts of fertilizer [24]. Biological nitrogen fixation is a crucial alternative source of nitrogen, and can be enhanced along with other integrated nutrient management strategies such as animal manure and other biosolids, and recycling the nutrients contained in crop residues [27].

Conservation tillage: changes in tillage practices have had a significant effect on shifting conventional cereal-based cropping systems to more diversified crop rotations that utilize pulses or oilseeds and that result in less soil disturbance. Long-standing patterns of monoculture cereal cropping resulted in pest and disease outbreaks and erosion, and fallowing led to increased soil salinity and loss of soil nitrogen and water. The nitrogen fixation capability of pulses is the greatest environmental benefit into cereal-fallow rotations thereby reducing fertilizer nitrogen requirements in the current and succeeding crop along with improved soil capacity to supply nitrogen. Farmers in Saskatchewan changed the tillage requirements by managing herbicide practices leading to reduced rates of applications.

Changes in tillage practices have been an important part of shifts from conventional cropping systems, based on grain production, to more diversified crop rotations utilizing pulses or oilseeds. Importantly, implementing conservation tillage practices has often involved introduction of pulses and oilseeds into grain-based crop rotations. Many studies have demonstrated the nitrogen fixation benefits of conservation- or no-tillage, with pulse and oilseed bean nodulation improving after multiple years of no-till and nitrogen fixation rates increasing (moderated by changes in rainfall patterns) [28].

Productivity vs. area expansions: Addition of pulses into different crop rotations increases the efficiency of any production system and hence increases the overall crop yields by reducing the need of expanded production area.

In Sub-Saharan Africa to improve the production rate for accomplishment of the food and feed demands is very crucial as maximum pulse production, although relatively low, occurs in rainfed areas with minimal inputs. It is estimated that the requirement of pulses (mainly cowpeas and beans) is expected to increase up to 155% till 2050. However, there are some examples which indicate enhanced production rates i.e., the yield of cowpea is increased at a greater proportion than the hectares planted in Ghana. This reflects the efficiency of production system due to supply of better quality seeds and improved varieties, cost-effectiveness for farmers, increased demands for better markets.

Climate change mitigation and adaptation: In crop rotation system, due to lower fertilizer requirement, pulses can aid in lowering GHG emissions. In Western Canadian cropping systems, about 70% of non-renewable energy used is due to nitrogen fertilizers specifically. SK, research at Swift Current, evaluated net GHG emission in four cropping systems (fallow-wheat-wheat, fallow-flax-wheat, lentil-wheat and continuous wheat). They observed the lentil-wheat system as the most efficient in GHG emissions due to the lower requirement of nitrogen fertilizer by wheat. In addition, increased nitrogen availability results into enhanced biomass production.

8. Social impacts

Nutrition and disease: the food consumption per capita may decline in developed countries by 2024, but in developing countries the demand will increase primarily based on increase in protein consumption. Globally, the contents of human diet shifted to more energy-dense foods containing highly saturated fats which are very risky for developing obesity, diet-related diseases like diabetes, coronary heart disease and cancer, etc. Along with the fruits and vegetables, pulses and legumes are important diet rich in micronutrients for healthy choices. Pulses help to control cardiovascular disease, increasing gut health and healthy nutrition. The global consumption level of pulses is declining, specifically in developing countries e.g., in 1963 pulse consumption in China was 30 g per capita per day which declined significantly to only 3 g per capita by 2003 [29].

The total caloric consumption is expected to increase at global level. The rate of food consumption in developed countries is in stagnant growth whereas it is significantly increasing in developing countries, as indicted by OECD/FAO Agricultural Outlook to 2024, reflecting increase in protein consumption. The rate of consumption of cereals is expected to increase about 390 Mt by 2024, which suggest that cereals will remain most consumed agricultural product [30]. India provides a counterpoint to China, as pulses there provide an increasing source of protein, now accounting for almost 13% of overall protein intake [31]. India is the largest pulse producer and consumer, and the country grows the largest varieties of pulses in the world, accounting for about 32% of the area and 26% of world production. The increased pulse crop yield from 0.63 t/ha in 2007–2008 to 0.79 t/ha in 2012–2013 is a measure of efficient production system. Along-with this, the annual yield growth is likely to overtake in production area. However, the production of pulses in India is still below the global average as the Indian needs for the pulses are increasing, causing more import, which is expected to grow about 5.1 Mt by 2023 [31].

Nutrition and food security: food security may be best defined as: "a situation comprising of all people always having access to safe, sufficient and nutritious food fulfilling their all food preferences and dietary needs to run a normal, active and healthy life in all good physical, social and economic environment [32]. Pulses add minerals, vitamin B, proteins as well as essential amino acid like lysine to cereals making it protein-rich. A number of new methods, in manufacturing food products, are being used now days to increase production and use of pulse protein fractions.

Gender: gender aspect of pulse production relates primarily to women's involvement in pulse production commercially, to feed families, and to benefit from income derived from pulse sales.

9. Economic impacts

Reduced reliance on fossil fuels and lower fuel costs: Pulses and oilseeds are commonly used in crop rotations where conservation of tillage practices has been adopted. By adopting no-till practices, the farmers are expected to see the long-standing economic benefits in terms of lesser degradation of soil, air and water with reduced consumption of fossil fuels.

10. Economic challenges for pulses in India

India is the largest consumer of pulses, but government subsidies and price controls in the agricultural sector created distortions that affected domestic production. In 1990s, government removed the import restrictions and lowered the tariffs on agricultural products which resulted into reversal of protectionist policies of 1970s and 1980s. The external trade in all major crops was regulated except basmati rice and durum wheat, and government agencies were authorized for all the imports. However, the import tariffs on pulses were considered differently which were gradually reduced and ultimately eliminated by 1996. The hope of augmented imports due to liberalization of domestic pulse market also could not be materialized rather; total pulse imports share in merchandise trade fell down after liberalization [33]. Minimum support prices are an important part of the policy decision to stabilize commodity prices, improve the economic viability of farming in India, this also enhance food security by diversification into oilseeds, pulses, livestock and fish. However, the MSP for pulses did not give the same trend as that of rice. During 2008–2009, prices for pulses were increased at a rate higher than for food grains, but the area covered under pulses not increased proportionally and this resulted into pulse cultivation risks. If we compare this case with paddy crop cultivation such risks are not associated, and the farmers are also assured for procurement by government agencies [31].

11. Pulses and livestock feed diversification

Pulses not only increase nitrogen supply [34] but also increase the meat production if used in livestock production systems. By 2050, demand for meat may increase to 200 MT per annum, with corresponding demand for livestock feed [35]. Field pea is being promoted as enriched source of protein and energy in cattle feed in the northern Great Plains of the US and Canada since it is easily digestible by cattle, but the starch fermentation and ruminal protein degradation rates are slower than for other common feeds. Dry matter intake by cows is also increased using field pea in livestock feed ration acting as a binding agent for pelleting formula feeds [36].

Significant economic benefits have been recorded in investment policies in pulse crop research. Across four CGIAR centers, the CGIAR Research Program on Grain Legumes, a global alliance coordinating efforts estimated that the net present value of gross benefits of its legume research and extension was US$ 4.5 billion in 2012, nearly to US$ 535 million per year. During 2014–2020, proposed activities by CGIAR program included legume research to address food security through increased availability of food (over 8 MT), nutrition security from more availability of protein, and environmental benefits through biological nitrogen fixation (a fertilizer cost saving of US$ 418 million). In South and South-East Asia and Sub-Saharan Africa, where most of the world's poorest communities are located, the CGIAR estimated more than 50% of the projected economic benefits through legume research and extension [37, 38].

12. Suitability of some pulses for marginal areas

Drought prone areas having lowest rainfall 300–450 mm/year are best suited for cultivation of drought-tolerant pulses including lentils, Bambara beans and pigeon peas. Normal crops cannot survive and produce under such adverse conditions. The benefit of these drought-tolerant legumes is their adaptability towards such harsh environments by deep-root systems surviving in water scare regions and thus improves the crop productivity in marginal environments. Using locally adapted pulses, farmers in dry environments, can intensify their production systems in a sustainable manner where food security represents a huge challenge. Additionally appropriate policies and programs, marketing trade systems to support the marginal farmers need to be encouraged for pulses to increase the availability and consumption of drought-resistant pulses.

Malnutrition is a major issue in many countries and pulses can be grown in these regions to overcome the hunger threat. The food and nutritional security can be achieved to some extent with the production of pulses in these regions [39].

Farmers are forced to use saline water for crop irrigation in arid and semi-arid regions due to water shortage or by poor quality water. Every crop has a threshold level of salinity tolerance above which yields decrease with increasing salinity. Soluble salts accumulate in saline soils affecting plant growth at various stages and creating yield differences of crops at maturity. This requires immediate controlling measures for soil salinization and preponderance of saline water sources along with cultivation of salt resistant crops.

In India, nearly 6.73 M ha area is affected with salinity and sodicity stresses covering various states of the country. Nearly 20% of the irrigated agricultural land is transforming into saline area with 1–3% per year either due to natural salinity or due to human interference. Global effects of increased salinity at agronomically important land will be visible by the middle of the twenty-first century [40–42]. Further, the arid and semiarid areas in different states are associated with saline underground water, which have to be used for irrigation purpose. The development and use of plant species that can tolerate high salt level is important for sustainable crop production on such soils and water conditions and is cost effective. This may be achieved by making use of variations in tolerance both, between and within cultivars. Low yield potential coupled with biotic and abiotic stresses has further reduced cultivation of grain legumes (chickpea and mung bean) by the farmers. Recently, realizing the significance of grain legumes in improving nutrition and the livelihood of poor farmers, more research is now being carried out for their genetic amelioration by various institutes. Though CSSRI have released a salinity tolerant desi chickpea variety (CSG 8962) in 1997, yet further improvements are required to recombine salt tolerance and high yield, which is the need of hour.

Omic approaches for crop improvement: the networks of genetic and environmental factors controlling various abiotic stresses are complex and hamper breeding strategies. There is a limit for traditional approaches for crop improvement; novel approaches in agriculture need to be adopted to meet the demands of an ever-growing world population. Various technological advances have led to the emergence of high throughput tools to explore and exploit plant genomes for crop improvement to counteract the aforementioned agricultural challenges. These approaches aim to explore the entire genomics to gain insights into plant molecular responses ultimately to provide specific strategies for crop improvement. Functional genomics techniques have long been adopted to unravel gene functions and the interactions between genes in regulatory networks, which can be exploited to generate improved varieties.

In these contexts, exploring management strategies to use low inorganic N with suitable grain legumes would help to sustain crop productivity.

Future strategies for increasing pulses productivity and production: to increase area and production of pulse crops we need crop specific and region specific approaches, which should be adopted in the overall framework of systems approach. The major thrust areas to be addressed are as follows.

- Input responsive and non-lodging varieties
- Biotic and abiotic stress tolerance
- Super-early varieties for green grains
- Machine harvestable and herbicide tolerant varieties
- Nutritionally enhanced varieties
- Integrated pest and disease management
- Public-private partnership for sustaining chain and to minimizing post-harvest losses
- Linking MSP to market prices can bridge the gap between demand and supply
- Climate smart varieties

13. Next steps in the pulse contribution

While the global pulse industry, and the pulse industry in India, successfully celebrated International Year of Pulses in 2016, let us also think about what will be done to ensure that pulses lead by example to help end hunger. The Zero Hunger Challenge and 2016 International Year of Pulses are opportunities to make a significant difference in the global fight against hunger, and to demonstrate to the world how pulses support healthy people and a healthy planet [38]. The aim of National Food Security Mission of India is to improve production of pulses and for this approximately Rs 1100 crore were distributed for during 2016–2017. The positivity of this program was executed by organizing quality seed production-cum-awareness field days highlighting the importance of quality seeds through allocation of Rs. 20.39 crore to ICAR/Agriculture Universities for increasing the availability of new pulse-variety breeder seeds. Inter cropping of pulses with other crops is being encouraged. A number of schemes have been launched for the development of agriculture and farmers' welfare. In view of good monsoon in India, in spite of two consecutive drought years, pulse production reached 25.23 MT (2017–2018) which is still lower than the domestic demand (27.91 MT) (Source: Directorate of Economics and Statistics, Department of Agriculture & Cooperation and Department of Commerce, Govt of India). Therefore, the BRICS nations have been approached and it was commented that India would like to seek cooperation from member countries (BRICS) in helping to meet our production shortfall in crops like pulses and oilseeds.

Let us show the same creativity and leadership that has built a strong global community within the pulse industry and also become the leaders of a basic human right—the Right to Adequate Food. It is time to get to work. It is time for the global pulse industry to step up to the challenge!

Author details

Anita Mann*, Ashwani Kumar, Satish Kumar Sanwal and Parbodh Chander Sharma
ICAR-Central Soil Salinity Research Institute, Karnal, Haryana

*Address all correspondence to: anitadgr13@gmail.com

References

[1] FAO (Food and Agriculture Organization of the United Nations). Definition and classification commodities, 4. Pulses and Derived Products. 1994. Available from: http://www.fao.org/es/faodef/fdef04e.htm

[2] Ali M, Gupta S. Carrying capacity of Indian agriculture: Pulse crops. Current Science. 2012;**102**(6):874-881

[3] Gowda Laxmipathi CL, Srinivasan S, Gaur PM, Saxena KB. Enhancing the productivity and production of pulses in India. In: Shetty PK, Ayyappan S, Swaminathan MS, editors. Climate Change and Sustainable Food Security. New Delhi: National Institute of Advanced Studies, Bangalore and Indian Council of Agricultural Research; 2013

[4] FAOSTAT. 2015. Available from: http://faostat.fao.org/site/339/default.aspx

[5] Lev-Yadun S, Gopher A, Abbo S. The cradle of agriculture. Science. 2000;**288**:1602-1603

[6] Cohen MN. The Food Crisis in Prehistory: Overpopulation and the Origins of Agriculture. New Haven, CT: Yale University Press; 1977

[7] Hymowitz T, Singh RJ. Taxonomy and speciation. In: Wilcox JR, editor. Soybeans: Improvements, Production and Uses. 2nd ed. Madison, WI, USA: American Society of Agronomy, Crop Science Society of America, and Soil Science Society of America; 1987. pp. 23-48

[8] Kaplan L, Lynch TF. Phaseolus (Fabaceae) in archaeology: AMS radiocarbon dates and their significance for pre-Columbian agriculture. Economic Botany. 1999;**53**:261-272

[9] Graham PH, Vance CP. Legumes: Importance and constraints to greater use. Plant Physiology. 2003;**131**:872-877

[10] Polhill RM, Raven PH, Stirton CH. Evolution and systematics of the Leguminosae. In: Polhill RM, Raven PH, editors. Advances in Legume Systematics, Part 1. Kew, UK: Royal Botanic Gardens; 1981. pp. 1-26

[11] Singh RP. Status Paper on Pulses. In: Government of India, Ministry of Agriculture (Department of Agriculture & Cooperation), Directorate of Pulses Development. 2013

[12] Saroja VN. 2016. Available from: http://Agriwatch.com

[13] Doering EJ, Sandoval FM. Hydrology of saline seeps in the northern Great Plains. Transactions of ASAE. 1976;**19**(856):861-865

[14] Sharma BR, Minhas PS. Strategies for managing saline/alkali waters for sustainable agricultural production in South Asia. Agricultural Water Management. 2005;**78**:136-151

[15] Mann A, Kaur G, Kumar A, Sanwal SK, Singh J, Sharma PC. Physiological response of chickpea (*Cicer arietinum* L.) at early seedling stage under salt stress conditions. Legume Research. 2019;**42**(5):625-632

[16] Sanwal SK, Kumar A, Mann A, Kaur G. Differential response of pea (*Pisum sativum*) genotypes exposed to salinity in relation to physiological and biochemical attributes. Indian Journal of Agricultural Sciences. 2018;**88**(1):149-156

[17] Vijay P, Choudhary Anil K, Anchal D, Bana RS, Rana KS, Rana DS, et al. Improved crop management practices for sustainable pulse production: An Indian perspective. Indian Journal of Agricultural Sciences. 2015;**85**(6):747-758

[18] Gan YT, Zentner RP, Campbell CA, Biederbeck VO, Selles F, Lemke R.

Conserving soil and water with sustainable cropping systems: Research in the semiarid Canadian prairies. In: Presentation to 12th ISCO Conference; Beijing, China. 2002

[19] Lupwayi NZ, Rice WA, Clayton GW. Soil microbial diversity and community structure under wheat as influenced by tillage and crop rotation. Soil Biology and Biochemistry. 1998;**30**:1733-1741

[20] UNEP. Growing Greenhouse Gas Emissions Due to Meat Production. Web. 2015

[21] Zentner RP, Walla D, Nagy C, Smith E, Young D, Miller P, et al. Economics of crop diversification and soil tillage opportunities in the Canadian prairies. Agronomy Journal. 2002;**94**:216-230

[22] Zentner RP, Campbell CA, Selles F, Lemke R, McConkey BG, Fernandez MR, et al. Economics of spring wheat production systems using conventional tillage management in the Brown soil zone – Revisited. Canadian Journal of Plant Science. 2007;**87**:27-40

[23] Campbelll CA, Lafond GP, Biederbeck VO, Winklemanr GE. Influence of legumes and fertilization on deep distribution of available phosphorus (Olsen-P) in a thin Black Chernozemic Soil. Canadian Journal of Soil Science. 1993;**73**:555-565

[24] Canfield D, Glazer A, Falkowski P. The evolution and future of Earth's nitrogen cycle. Science. 2010:330

[25] Smil V. Nitrogen and food production: Protein for human diets. Ambio. 2002;**31**(2):126-131

[26] Peoples M, Swan T, Goward L, Hunt J, Li G, Harris R, et al. Legume Effects on Soil N Dynamics—Comparisons of Crop Response to Legume and Fertiliser N. Government of Australia: Grains Research and Development Corporation; 2015. Available from: http://grdc.com.au/Research-and-Development/GRDC-Update-apers/2015/02/

[27] Lal R. Carbon emissions from farm operations. Environment International. 2004;**30**:981-990

[28] van Kessel C, Hartley C. Agricultural management of grain legumes: Has it led to an increase in nitrogen fixation? Field Crops Research. 2000;**65**:165-181

[29] Kearney J. Food consumption trends and drivers. Philosophical Transactions of the Royal Society B. 2010;**365**:2793-2807

[30] OECD-FAO. 2015. OECD-FAO Agricultural Outlook 2015-2024

[31] OECD/FAO. 2014. OECD-FAO Agricultural Outlook 2014-2023

[32] FAO. The State of Food Insecurity in the World 2001. Rome: Food and Agriculture Organization of the United Nations; 2001

[33] Agbola F. Agricultural policy reform in India: Implications for pulse trade, prices and production, 1970-1999. Australasian Agribusiness Journal. Agribusiness Perspectives Papers, Paper 63. 2004

[34] Lemke RL, Zhong Z, Campbell CA, Zentner R. Can pulse crops play a role in mitigating greenhouse gases from North American agriculture. Agronomy Journal. 2007;**99**:1719-1725

[35] Alexandratos N, Bruinsma J. World agriculture towards 2030/2050: The 2012 revision. In: ESA Working Paper No. 12-03. Rome: FAO. Available from: http://www.fao.org/3/a-ap106e.pdf

[36] CGIAR. 2012. Available from: www.cgiar.org

[37] Kissinger G. Lexeme Consulting. Pulse Crops and Sustainability: A

Framework to Evaluate Multiple Benefits. Available from: http://www.FAO.ORG/pulses-2016/

[38] FAO. Pulses Contribute to Food Security. 2016. Avaialble from: fao.org/pulses-2016

[39] Kandil AA et al. Genotypic differences between two Mungbean varieties in response to salt stress at seedling stage. International Journal of Agriculture Sciences. 2012;**4**(7):278-283

[40] Karthikeyan A, Valarmathi R, Nandini S, Nandhakumar MR. Genetically modified crops: Insect resistance. Biotechnology. 2012;**11**:119-126

[41] Karuppanapandian T, Karuppudurai T, Sinha PB, Kamarul HA, Manoharan K. Genetic diversity in green gram (*Vigna radiata* L.) landraces analyzed by using random amplified polymorphic DNA (RAPD). African. Journal. Biotechnology. 2006;**5**:1214-1219

[42] Gordon. Pulse Canada. 2016. Available from: http://www.fao.org/righttofood/about-right-to-food/en/

Chapter 7

Helping Legumes under Stress Situations: Inoculation with Beneficial Microorganisms

Salvadora Navarro-Torre, Khouloud Bessadok,
Noris J. Flores-Duarte, Ignacio D. Rodríguez-Llorente,
Miguel A. Caviedes and Eloísa Pajuelo

Abstract

In the upcoming years, legume crops will be subjected to multiple, diverse, and overlapping environmental stressors (raise in global temperatures and CO_2, drought, salinity, and soil pollution). These factors will menace legume productivity and food quality and security. In this context, tolerant plant growth promoting rhizobacteria (PGPR) are useful biotechnological tools to assist legume establishment and growth. In this chapter, tolerant PGPR able to promote legume growth will be revised. Besides, in the era of -omics, the mechanisms underlying this interaction are being deciphered, particularly transcriptomic, proteomic, and metabolomic changes modulated by PGPR, as well as the molecular dialog legume-rhizobacteria.

Keywords: PGPR, legumes, abiotic stress, mechanisms, bacteria

1. Introduction

Plants are colonized by several microorganisms, mainly bacteria, and the number of them sometimes is higher than the number of plant cells [1]. These microorganisms can live inside (endophytes) and outside (epiphytes) the plant tissues, and they do not cause diseases in the host plant [2]; rather, these microorganisms contribute to improve the health and the productivity of the plants [3].

Soils also have many microorganisms with potential to improve plant growth, and overall the rhizosphere. The rhizosphere is an area of interaction between microorganisms and plant roots, and it is inhabited by bacteria, fungi, protozoa, actinomycetes, and algae, with bacteria and mycorrhizal fungi being the main populations [4]. The size of the microbial population in the rhizosphere of plants is influenced by root exudates. The chemicals found in the soil along with the exudates from the plants cause changes in the pH and in the redox potentials that will be determined by the microbial community around the roots [5].

Among the bacteria that colonize the rhizosphere, those promoting plant growth, also known as PGPB (Plant Growth Promoting Bacteria), rhizobacteria or PGPR (Plant Growth Promoting Rhizobacteria), and nitrogen-fixing rhizobia, are the most remarkable because they provide beneficial effects in the development of the plants being able to do it by direct or indirect mechanisms. Regarding legumes, they also interact with rhizobacteria, and the main interaction is the symbiosis between legumes and nitrogen-fixing rhizobia [4]. Rhizobia are bacteria that belong to the family Rhizobiaceae, and the most important genera are *Bradyrhizobium*, *Ensifer*, *Mesorhizobium*, *Rhizobium*, and *Sinorhizobium* [6]. During this interaction, rhizobia invade the root and group into a specialized organ called nodule. Inside the nodules, rhizobia become bacteroids, which transform N_2 to NH_4^+, molecule that can be assimilated by plants [7]. Inoculation of legumes with rhizobacteria produces a bioaugmentation of the microbial population in soils, thus contributing to plants growing bigger and faster [8–10], and this can solve the problem of the rapid growth of the world population that causes a great pressure in the area of land destined for food [11]. Moreover, the arable lands are decreasing because of the climate change and some human activities.

Climate change is one of the most important problems in the planet nowadays. Because of that, temperature and drought are increasing, involving an increase in arid and semi-arid zones and generating a loss of arable soils [12–14]. Drought is an abiotic stress that causes the highest losses in agriculture, so it is a very important factor in crop productivity [15]. Drought and heat involve the appearance of saline soils [16], although some human activities, like the increase of irrigation with bad water quality, are also responsible [17]. Salinity affects around 800 million of hectares in the world, and it is considered a global problem [18] being another stress that limits plant growth, productivity, nitrogen fixation in legumes, and the seed germination [19, 20] due to the uptake excess of NaCl by the plants [21]. Furthermore, the salt excess decreases the organic matter in soils and modifies the microbial population in the rhizosphere [22], so salinity also affects the nodulation negatively [23]. Finally, an additional abiotic stress is heavy metals. The increase of heavy metal concentrations becomes a pollution problem, being humans the main responsible of it [24]. Heavy metals affect plants and soils as the rest of abiotic stress does, and in legumes, they dramatically reduce nodule number and nitrogen fixation [25, 26].

To try to recover these affected soils, phytoremediation is being used, and several studies confirm that it is a very efficient tool, particularly in combination with bacterial inoculation since PGPB improve the potential of plants to phytoremediate soils [27–31]. Legumes belong to the plants used in phytoremediation because this family is one of the most diverse among other plant families in the world, and some of them are able to grow in degraded soils and can be used as pioneer plants in order to repopulate these degraded areas [25, 32]. In fact, legumes are usually used in intercropping with other crops to decrease the amount of pesticides and improve the quality of soils making legumes one of the most promising components of the Climate Smart Agriculture concept [33]. As described above, the symbiosis of legumes with rhizobia improves the growth of legumes and allows them to grow better in the degraded soils, but all the named abiotic stresses interfere with this interaction. For that, authors look for rhizobia resistant to these stresses that able to grow and form nodules even in degraded soils [34–37]. Furthermore, several studies demonstrate that coinoculation of legumes with rhizobia and another PGPR increases nodulation, plant growth, and the potential to phytoremediate soils of plants in the presence of abiotic stresses [38–41]. This improvement in legumes occurs for the interaction between plants and bacteria through different direct and indirect mechanisms that help the plant to grow in the presence of stress.

In this chapter, the different bacterial mechanisms used to improve the plant growth in the presence of the most important abiotic stresses nowadays are been reviewed, in order to help legumes to grow under stress situations and recover the degraded soils using the interaction between legumes and bacteria. Furthermore, the molecular mechanisms involved in these interactions are being described with the transcriptomic, proteomic, and metabolomic studies so far.

2. Effects and mechanisms of plant growth promotion by microorganisms

As mentioned above, some bacteria are capable of promoting plant growth (PGPB and PGPR) through direct (biofertilization) and indirect (biocontrol) mechanisms. The direct mechanisms are based on the direct promotion of plant development, among which are nitrogen fixation, phosphate solubilization, production of phytohormones (auxins, cytokinins, and gibberellins), the enzymatic activity of the aminocyclopropane carboxylic acid (ACC) deaminase, and iron complexation by bacterial siderophores. On the other hand, indirect methods are responsible for inhibiting pathogenic organisms for plants. Among these methods are the synthesis of antibiotics, enzymes that degrade the cell wall, or the induced systemic resistance (ISR) process [42]. The mechanisms carried out by PGPB and PGPR will depend on the host plant and will be influenced in turn by biotic (such as plant defense mechanisms and genotype) and abiotic (such as weather conditions and soil composition) factors [43].

However, bacteria are not the only microorganisms that are able to promote plant growth. Mycorrhizal fungi also can carry out this function, there being a symbiotic association between them and most terrestrial plants [44] favoring micronutrient absorption, resistance to diseases caused by pathogens, or reduction of plant stress caused by environmental factors [45].

2.1 Direct mechanisms

2.1.1 Fixation of nitrogen

Nitrogen is one of the essential elements for life that is present in biochemical structures as important as nucleotides and proteins, but atmospheric nitrogen is mostly nonassimilable for plants since about 78% is in a gaseous state, so it must become ammonia, thanks to the nitrogenase enzyme to be assimilable. This reaction is carried out by rhizobia under symbiosis with legumes, thanks to which the rhizobia obtain carbon provided by the legumes from photosynthesis, and they provide the plant with nitrogen [46].

Atmospheric nitrogen fixation occurs mainly in leguminous plants where rhizobia/plant interactions are highly specific [42]. However, certain free-living bacteria (such as *Frankia* spp. or Actinobacteria) are also able to fix atmospheric nitrogen to a much wider range of plants than rhizobia [47]. For example, coinoculation of *Bradyrhizobium* sp. UFLA 03-84 with *Actinomadura* sp. 183-EL, *Bacillus* sp. IPACC11, or *Streptomyces* sp. 212 in cowpea plants improves the nitrogen fixation even in the presence of salinity conditions [48]. Another example of the improvement of nitrogen fixation in the presence of a different abiotic stress is reported by Saia et al. [49] in which plants of *Trifolium alexandrinum* were inoculated with arbuscular mycorrhizal (AM) fungi leading to an improvement of nitrogen fixation and the plant growth in water stress.

2.1.2 Solubilization of phosphate

Phosphorus is also an essential element for life that is involved in such important processes as energy transfer, respiration, or photosynthesis [50]. After nitrogen, phosphorus is the second most limiting element for plant growth [51]. The content of this element in the soil is 0.05% (w/w) of which only 0.1% can be used by plants, a problem to which the use of phosphorus-based chemical fertilizers that are fixed within the soil must be added and limited the bioavailability of the element [52].

There are a wide variety of microorganisms that can solubilize phosphate to make it assimilable to plants by releasing phosphorus from soil insoluble phosphates. An example of this is the endophytic bacterium *Gluconacetobacter diazotrophicus* that is capable of solubilizing phosphate by acidification [53]. Bacteria also can solubilize phosphates in the presence of different stresses such as *Serratia* sp. J260, *Pantoea* sp. J49, *Acinetobacter* sp., L176, and *Planomicrobium* sp. MSSA-10 that kept this property in the presence of salinity conditions or *Bacillus* sp. L55 in the presence of high temperatures [54, 55]. Besides, other bacteria are able to solubilize organic phosphorous because they produce phytase and phosphatases enzymes that act by dephosphorylating phytates and organic phosphorus [51]. This kind of bacteria can help to legume crops to improve their growth and productivity in these abiotic stress conditions.

2.1.3 Phytohormones production

There are certain microorganisms capable of producing phytohormones, which can promote or modify the development of plants at low concentrations [43]. Among the most common phytohormones are gibberellins, cytokinins, auxins, ethylene, and abscisic acid [53], and their effects can be stimulants or inhibitors of plant growth. The most widely studied are auxins, particularly the indole-3-acetic acid (IAA), being the one that is produced in a greater proportion by plants and PGPR where up to 80% of rhizobacteria synthesize it as a secondary metabolite [56]. This auxin is present in different cellular responses such as cell division, gene expression, or root development and lengthening and affects photosynthesis, pigment formation, and resistance to stress conditions [53].

When the concentration of IAA in plants is adequate, the application of bacterial IAA can have positive, negative, or neutral effects [46], so that this bacterial synthesis will determine the stimulation or inhibition of plant growth and may change the hormone level from optimal to supraoptimal. In this way, the PGPR will stimulate growth when the IAA concentration is below the optimum levels for the plant [42]. It is important that IAA producing bacteria must keep the property even in the presence of any abiotic stress to help plants to grow in these conditions. **Table 1** shows some examples of IAA producing microorganisms that cause improvements in legume crops under stress conditions.

2.1.4 ACC deaminase activity

Ethylene, also known as the stress hormone, is a phytohormone present in all higher plants, making it a key element in a wide range of biological activities, intervening in processes that affect the growth and development of plants where almost all plant tissues and their stages of development are affected by it. It is produced from 1-aminocyclopropane-1-carboxylic acid (ACC) and is catalyzed by the enzyme ACC oxidase [59, 60]. Most abiotic stresses cause a large increase in ethylene concentration causing wilting of the flowers or initiating senescence in the leaves among other consequences, so its increase translates into harmful effects on the growth and health of plants and therefore to a reduction in crop production. However, when its concentration is adequate, it decreases the wilting, stimulates the germination of seeds, and influences the fruit ripening [61]. More recently,

Microorganism	Plant host	Plant improvement	Abiotic stress	References
Ochrobactrum cytisi Azn6.2	*Medicago sativa*	Larger root size	Heavy metals	[37]
Pseudomonas extremorientalis TSAU20 and *P. trivialis* 3Re27	*Galega officinalis*	Improved the growth and increased the nodulation	Salinity	[57]
Bradyrhizobium sp. RM8	*Vigna radiata*	Reduced the uptake of Ni and Zn and increased the nodulation	Heavy metals	[58]
Ensifer meliloti RD64	*M. sativa*	Accumulation of osmoprotectants, greater Rubisco availability	Drought	[8]
Bradyrhizobium sp. RJS9–2	*Stylosanthes guianensis*	Higher salt tolerance and osmoprotectants	Salinity	[9]
Aspergillus japonicus EuR-26	*Glycine max*	Improved the plant biomass and other growth features	Heat	[10]

Table 1.
Beneficial interactions between IAA producing bacteria and legumes under stress conditions.

evidence is being presented showing that ACC deaminase producing microorganisms can facilitate nodulation of legumes under stress conditions, called as nodulating helper bacteria [62, 63].

The great importance of this hormone is the reason for it having being extensively studied, so a wide range of microorganisms capable of secreting the enzyme ACC deaminase is known. ACC deaminase hydrolyses ACC in ammonia and 2-oxobutanoate, thus causing a reduction in the concentration of ethylene in plants [64]. Among ACC deaminase producing microorganisms are the following genera: *Bacillus*, *Pseudomonas*, *Klebsiella*, *Serratia*, *Arthrobacter*, *Azospirillum*, *Streptomyces*, *Microbacterium*, *Achromobacter*, *Acinetobacter*, *Acidovorax*, *Agrobacterium*, *Alcaligenes*, *Enterobacter*, *Agrobacterium*, and *Rhizobium* [59, 60]. The interaction of these microorganisms with legumes enhances plant growth and crop productivity under stress conditions (**Table 2**).

2.1.5 Production of siderophores

Iron is another essential micronutrient for plants that, in the oxygen-rich conditions of the rhizosphere, is in the form of Fe^{+3} that is insoluble for plants and microorganisms [53]. Siderophores, involved in both direct and indirect mechanisms of plant growth promotion, are small molecules of a peptide nature formed by side chains and functional groups that act as ligands with high affinity for the Fe^{+3} ions [65]. A wide range of bacteria and fungi are capable of secreting siderophores that occur in rhizospheric soils in neutral-alkaline pH conditions, where there is a deficiency of this element due to its low solubility in these conditions [66]. These microorganisms can subsequently absorb the siderophore-Fe^{+3} complex by means of a specific receptor and release the Fe in its bioavailable form (Fe^{+2}) to support bacterial growth [61]. The creation of this complex also assumes an important role in the adsorption of Fe by plants, in the presence of other metals such as nickel or cadmium [43]. Another very important function of siderophores is to prevent the proliferation of pathogens by competing for the available iron. In this way, rhizobacteria help plant growth by releasing these biocontrol agents against phytopathogens (antagonism of PGPR against pathogens) [61].

Microorganism	Plant host	Abiotic stress	Benefits for plants	References
Pseudomonas fluorescens TDK1	*Arachis hypogea*	Salinity	Improved plant growth parameters and alleviated saline stress	[67]
Pseudomonas aeruginosa GGRJ21	*V. radiata*	Drought	Increased plant biomass, the relative water content, and osmolytes	[68]
Bacillus subtilis LDR2	*Trigonella foenum-graecum*	Drought	Alleviated ethylene-induced damage and improved nodulation and mycorrhizal fungi colonization	[69]
Arthrobacter protophormiae SA3	*Pisum sativum*	Salinity	Alleviated ethylene-induced damage and improved nodulation and mycorrhizal fungi colonization	[70]
Ochrobactrum pseudogrignonense RJ12, *Pseudomonas* sp. RJ15 and *B. subtilis* RJ46	*Vigna mungo* and *P. sativum*	Drought	Increased seed germination, plant biomass, chlorophyll, and relative water content.	[71]

Table 2.
Beneficial interactions between ACC deaminase producing bacteria and legumes under stress conditions.

Microorganism	Plant host	Plant improvement	Abiotic stress	References
Bradyrhizobium sp. RM8	*V. radiata*	Reduction of Ni and Zn concentrations in plant tissues	Heavy metals	[58]
Bradyrhizobium sp. YL6	*G. max*	Positive effects in photosynthesis and mineral nutrients	Heavy metals	[72]
P. aeruginosa GS-33	*G. max*	Improve plant biomass, chlorophyll content, and reduction of fungal infections	Salinity	[73]
Pseudomonas putida NBRIRA and *Bacillus amyloliquefaciens* NBRISN13	*Cicer arietinum*	Higher growth and yield	Drought	[74]

Table 3.
Beneficial interactions between bacteria that produce siderophores and legumes and plant improvements under stress conditions.

In general, plants can benefit in many ways, thanks to the production of siderophores since they are also involved in the improvement of nitrogen fixation or in the prevention of toxicity by heavy metals [75]. Thus, PGPR able to produce siderophores have a certain competitive advantage over other microorganisms in the rhizosphere [64]. Some of these microorganisms are shown in **Table 3** showing the repercussions caused by the synthesis of iron chelating compounds depending on the crop and the conditions of the plant-microbe interaction.

2.2 Indirect mechanisms

Indirect mechanisms are those processes where PGPR prevent or neutralize the harmful action of phytopathogens by the production of substances that confer greater natural resistance to the host plant, protecting it against infections (biotic stress), although they also help plants to grow actively under conditions of environmental stress (abiotic stress) [43].

2.2.1 Production of antibiotics

Antibiotic production is the main mechanism by which a large and heterogeneous group of bacteria fight the harmful effects of plant pathogens (usually fungi). The antibiotics produced by PGPR are low molecular weight compounds that negatively interfere with the metabolic processes of other microorganisms, thus delaying their growth [64] or inhibiting it [56]. There are some examples of PGPR that produce antibiotics against phytopathogens reflected in **Table 4**.

The effectiveness with which these molecules interfere with pathogen suppression will depend on the metabolite secreted by the PGPR and environmental conditions (mineral content of the soil, osmotic conditions, carbon sources, etc.) [76]. Moreover, some phytopathogens may develop resistance to specific antibiotics by repeated use of the same strain that produces a particular antibiotic, so it is preferable to inoculate plants with PGPR that produce several antibiotics [59]. There are some PGPR that have antagonistic activities against some phytopathogens in addition to improve plant growth in the presence of some stress as it is the case of *Cellulosimicrobium funkei* AR6 that improves the root elongation in crops of *P. vulgare*, *V. radiata,* and *V. mungo* in the presence of Cr(VI) and also has a strong antagonistic activity against *Aspergillus niger* [77]. Another example is *B. thuringiensis* UFGS2 that improves plant growth, physiologic parameters, and the resistance of the soybean against *S. sclerotiorum* under drought stress [78].

2.2.2 Lytic enzymes of the cell wall

Some PGPR produce enzymes that are involved in the lysis of cell walls and neutralization of pathogens by interrupting a particular stage of development or the cell cycle [79], playing an important role in promoting plant growth by protecting them of biotic and abiotic stresses due to the suppression of these pathogens. Among the produced enzymes for this purpose are chitinases, dehydrogenases, β-glucanases, lipases, phosphatases, or proteases [59]. The cell wall of most fungi is formed by residues of β-1,4-N-acetyl-glucoseamine and chitin, so that the bacteria that produce β-1,3-glucanase and chitinase can control the growth of phytopathogen [43]. Furthermore, some PGPR are able to produce this kind of enzymes and protect the crops under abiotic stress like *Bacillus licheniformis* A2 that produces

Microorganism	Plant host	Phytopathogen	Reference
Rhizobium sp. RS12	*C. arietinum*	*Fusarium solani* and *Macrophomina phaseolina*	[80]
Pseudomonas sp. YL23	*G. max*	*Erwinia amylovora* and *Dickeya chrysanthemi*	[81]
Pantoea ananatis RM_2	*P. sativum*	*Trichoderma longibrachiatum* and *Fusarium oxysporum*	[82]
Bacillus sp. B19, *Bacillus* sp. P12, and *B. amyloliquefaciens* B14	*Phaseolus vulgaris*	*Sclerotinia sclerotiorum*	[83]

Table 4.
PGPR that produce antibiotics and their effects as biocontrol agents.

chitinase and protects *A. hypogea* against *Fusarium oxysporum* f. sp. *cubense* under salinity conditions [84]. **Table 5** shows some examples of bacteria capable of producing these types of degrading enzymes.

2.2.3 Induced systemic resistance

There is an important feature of biocontrol that helps plant growth based on two defensive response mechanisms against various external agents known as mechanisms of induced systemic resistance (ISR) and mechanism of acquired systemic resistance (ASR) [56]. ISR is a physiological state of defensive capacity that plants present in response to an environmental stimulus [43] in which nonpathogenic microorganisms, including various PGPB, reduce the negative effects of pathogens of plants by activating a resistance mechanism without the need to target a specific pathogen and can develop this resistance in response to infection by pathogens, to attack by insects, or to a chemical treatment [42].

To stimulate this defensive response mechanism, the ISR uses phytohormones as jasmonic acid (JA) and ethylene (ET) that act as signaling molecules [64], although it has been shown that some organelles (such as flagella) and bacterial molecules (such as lipopolysaccharides of the outer membrane or antibiotics produced by bacteria) can also act as inducing agents activating ISR and generating a rapid accumulation of pathogenesis-related enzymes such as chitinase, β-1,3-glucanase, peroxidase, or liases, among others [85]. It is important to note that the ISR prepares plant to fight the pathogen with an improved defense [56]. A clear example of the ISR elicitor is the effect of *Bradyrhizobium japonicum* in soybean crops, where systemic redox changes are induced in plants [86]. The induction of ISR by *Bacillus* sp. CHEP5 and *Bradyrhizobium* sp. SEMIA 6144 in peanut plants against *S. rolfsii* also has been demonstrated [87]. However, this microbial induction could be limited by abiotic stress like the ISR induction of *B. amyloliquefaciens* S499 in different crops under heat and drought conditions where the response against *Botrytis cinerea* infection was prejudiced [88].

2.2.4 Quorum sensing

Quorum sensing (QS) is a mechanism of genetic regulation in response to cell density mediated by small self-inducing molecules, which are usually secreted out of bacterial cells and act as chemical signals produced by an increase in the cell population. These molecules cause an alteration in bacterial metabolism by activating different sets of genes [89], so that similar bacteria that live in communities and are close to each other will begin to act in a coordinated way. The level of the autoinductors increases proportionally to the population of bacterial cells until it

Microorganism	Host plant	Enzyme	Phytopathogen	References
B. licheniformis A2	*A. hypogea*	Chitinase	*F. oxysporum* f. sp. *cubense*	[84]
Bacillus altitudinis BRHS/S-73	*V. radiata*	Chitinase and protease	*Thanatephorus cucumeris*	[90]
B. subtilis DSM1088, *P. fluorescens* ATCC13525 and *Glommus* spp.	*P. vulgaris*	Chitinase, peroxidase and polyphenol oxidase	*Sclerotium rolfsii*	[91]

Table 5.
Lytic enzymes produced by PGPR and their effect on biocontrol of legume phytopathogens.

exceeds a defined threshold level (*quorum*) where it binds to bacterial cell receptors and triggers a signal transduction cascade leading to changes in the expression of bacterial genes by the action of a group of cells [42].

There are numerous bacterial signaling molecules among which the acyl-homoserine lactone (AHL) produced by Gram-negative bacteria consisting of a common lactone homo resin residue and an acyl chain that can vary between 4 and 18 carbon atoms. There are more than 100 species of proteobacteria that produce AHL, and three types of enzymes are known (LuxI, LuxM, and HdtS) capable of synthesizing AHL *in vivo.* Degradation of AHL can be carried out by various enzymes as is the case of the AHL lactonase that breaks the ester linkage of the lactose ring of the AHL to form N-acyl homoserine reversibly [92]. Most of these signal molecules are of the bacterial origin: *Bacillus* spp., *Klebsiella pneumoniae* KTCTC2241, or *Rhizobium* sp. NGR23 [93]. The detection of this bacterial communication system is interrupted by PGPB, thus preventing pathogen infection through immune responses and preventing plant growth inhibition [94].

3. Molecular mechanisms in the bacteria-plant interactions under abiotic stress

In spite of the abundant literature available on the improvement of plant growth and crop productivity using bacteria under abiotic stress conditions, the molecular mechanisms involved in these interactions remain elusive so far. This area has been studied by some authors to try to understand what changes bacteria elicit on plant gene expression finally leading to enhance the plant resilience to environmental stress. There are not many studies about genes involved in the plant-bacteria interactions, particularly in legumes, but in this section, some examples of these studies are exposed.

Plants have several mechanisms to adapt in the presence of any stress, whose genes are involved in the regulation of transporters, phytochelatins, glutathione reductase, phytohormones, oxidative stress reduction, phenolic compound, osmolytes, and low molecular weight organic acids, among others, and the bacteria role either expressing or repressing these genes has been investigated using real-time quantitative PCR, RNA Seq, and metabolomic and/or transcriptomic analyses. *OsGRAM* family genes are some of the most important in the plant growth and development under stress conditions, and it was demonstrated in rice under an array of stress situations, including salt. The plants inoculated with *B. amyloliquefaciens* SN13 showed overexpression of these genes [95].

Regarding heavy metal stress, one of the main genes expressed or repressed in plants is genes of transporters [96]. In the case of cadmium, *Tatm20* gene codifies a transmembrane transport that is expressed in the presence of Cd in wheat. When wheat plants were inoculated with *Azospirillum brasilense*, this gene was expressed much less than in the absence of the inoculation suggesting that *A. brasilense* helps plants to decrease Cd uptake and accumulation [97]. In this study, we also studied the expression of *TasSOS1* gene involving in keeping the plant cell homeostasis in the presence of high salt concentrations. In contrast to *Tatm20* gene, *TasSOS1* is overexpressed in inoculated plants under salinity stress. This overexpression maintains the homeostasis in plants and makes plant more tolerant to salt excess [97]. In *M. sativa* plants grown under heavy metal conditions, the genes involved in the phytochelatins synthesis were overexpressed in inoculated plants helping plant to detoxify and to grow in the presence of this stress. Moreover, the expression of glutathione reductase was improved by bacteria, so inoculated plants keep the redox status under heavy metal conditions [41]. Other examples are *Bacillus altitudinis*

FD48 that modules the expression of some genes involved in the synthesis of auxins improving root elongation in rice [98] and *Pseudomonas aeruginosa* and *Burkholderia gladioli* that reduced the expression of *CAT*, *GR*, *GPOX*, *APOX*, and *GST* genes in the presence of Cd stress in *Lycopersicon esculentum* plants modulating their antioxidative response [98]. *P. aeruginosa* and *Bacillus gladioli* also elevated the expression of phenolic compound, osmolytes, and molecular weight organic acid (citric acid, malic acid, fumaric acid, and succinic acid) genes, decreasing the physiological damage of *Solanum lycopersicum* plants under Cd toxicity [98].

A global transcriptomic analysis was carried in inoculated *Medicago truncatula* plants grown in the presence of arsenic and inoculated with an arsenic-resistant rhizobial strain. The results showed the overexpression of some defense genes in the inoculated plants (sulfur metabolism, several enzymes of the phytochelatins synthesis pathway, proline, heat shock proteins, and several transcription factors). Besides, secondary metabolism, isoflavonoids and phenylpropanoids, were activated. In contrast, the genes of nodulation were downregulated, particularly those involved in the early stages of the interaction [99, 100].

Under drought environments, *P. putida* MTCC5279 modulates the stress in plants of *Cicer arietinum* by the overexpression of *ACO* and *ACS* (involved in the synthesis of ethylene); *PR1* (synthesis of salicylic acid); *MYC2* (synthesis of jasmonate); *SOD, CAT, APX*, and *GST* (genes that codify the antioxidative enzymes in plants); *DREB1A* (response element to dehydration); *LEA* and *DHN* (dehydrins); and *NAC1* (transcription factors expressed under abiotic stress) genes [101]. In plants of sorghum inoculated with rhizobacterial endophytes, proline (a crucial molecule of maintaining the cellular functions under drought) accumulation was higher than in the noninoculated plants because bacteria induce the overexpression of *sbP5CS2* (pyrroline-5-carboxylate synthase 2) and *sbP5CS1* (pyrroline-5-carboxylate synthase 1) genes under drought stress [102]. This fact also was demonstrated in inoculated chickpea plants, where proline was accumulated by plants under drought conditions besides histidine, citrulline, and threonine [103].

Finally, the molecular mechanisms for plant alleviation in salt stress by bacteria are also reported by some authors. A transcriptome of rice plants inoculated with *Bacillus amyloliquefaciens* SN13 showed that the bacterial inoculation alters gene expression under salt stress. For example, genes of phytohormones, flavonoids, or photosynthesis are found in higher number in inoculated plants [104]. Other example is the inoculation of wheat with *Dietzia natrolimnaea* that causes an overexpression of genes involved in the ABA signaling cascade and in the salt sensitive pathway among others [105].

Recently, some authors have studied the miRNAs induced by PGPR as a possibility to regulate the stress in plants [106, 107]. miRNAs are RNA molecules of 20–24 nucleotides that do not codify proteins, and they get bound to mRNA or any transcriptional factor, regulating the expression of the target gene. However, the only study about miRNAs was performed in chickpea plants under drought stress, where plants were inoculated with *Pseudomonas putida* RA, and this inoculation improves plants adaptation to drought conditions through the regulations of miRNA expression [108].

4. Conclusions and future perspectives

One of the main problems of the rapid increase in the world's population lies in the challenge of having the necessary food for global supply, but the climate change and the pollution decrease the number of the agricultural soil, so a possible solution would be to encourage more widespread use of PGPB. The evidence that PGPR promote the plant growth under stress environments is becoming increasingly true, being even more focused on the study of individual mechanisms than their combined mechanisms, which is why new paths are being opened toward the use of mixed inoculants that act jointly acquiring faster and improving results. There are also investigations in the area of genetics and molecular biology, where studies are being carried out based on the discovery of specific genes capable of motivating greater plant development as well as in the field of nanotechnology where nanoencapsulation and microencapsulation offer an alternative to produce effective formulations against pest control. However, this area needs more investigations and funding to solve the lack of development of new and better storage, shipping, formulation and application techniques of these PGPR, and the development of effective and consistent regulations regarding their use. In this way, the agricultural practice in degraded soils could become an effective and sustainable practice for the benefit of all.

Acknowledgements

This research was funded by Junta de Andalucia (Spain) under FEDER 2020/00000092 project.

Conflict of interest

The authors declare no conflict of interest.

Author details

Salvadora Navarro-Torre[1]*, Khouloud Bessadok[1,2], Noris J. Flores-Duarte[1], Ignacio D. Rodríguez-Llorente[1], Miguel A. Caviedes[1] and Eloísa Pajuelo[1]

1 Departamento de Microbiología y Parasitología, Facultad de Farmacia, Universidad de Sevilla, Sevilla, Spain

2 Faculty of Sciences of Gabès, Research Unit: Valuation of Biodiversity and Bioresources in Dry lands (BVBZA), Gabès, Tunisia

*Address all correspondence to: snavarro1@us.es

References

[1] Mendes R, Garbeva P, Raaijmakers JM. The rhizosphere microbiome: Significance of plant beneficial, plant pathogenic, and human pathogenic microorganisms. FEMS Microbiology Reviews. 2013;**37**:634-663. DOI: 10.1111/1574-6976.12028

[2] Kuklinsky-Sobral J, Araújo WL, Mendes R, Geraldi IO, Pizzirani-Kleiner AA, Azevedo JL. Isolation and characterization of soybean-associated bacteria and their potential for plant growth promotion. Environmental Microbiology. 2004;**6**:1244-1251. DOI: 10.1111/j.1462-2920.2004.00658.x

[3] Berg G, Grube M, Schloter M, Smalla K. Unraveling the plant microbiome: Looking back and future perspectives. Frontiers in Microbiology. 2014;**5**:148. DOI: 10.3389/fmicb.2014.00148

[4] Nadeem SM, Ahmad M, Zahi ZA, Javaid A, Ashraf M. The role of mycorrhizae and plant growth promoting rhizobacteria (PGPR) in improving crop productivity under stressful environments. Biotechnology Advances. 2014;**32**:429-448. DOI: 10.1016/j.biotechadv.2013.12.005

[5] Timmusk S, Behers L, Muthoni J, Muraya A, Aronsson AC. Perspectives and challenges of microbial application for crop improvement. Frontiers in Plant Science. 2017;**8**:1-10. DOI: 10.3389/fpls.2017.00049

[6] Lloret L, Martínez-Romero E. Evolution and phylogeny of rhizobia. Revista Latinoamericana de Microbiología. 2005;**47**:43-60

[7] Sprent JI. Legume Nodulation: A Global Perspective. Hoboken: John Wiley & Sons; 2009. ISBN: 1444316397

[8] Defez R, Andreozzi A, Dickinson M, Charlton L, Tadini L, Pesaresi P, et al. Improved drought stress response in alfalfa plants nodulated by an IAA over-producing *Rhizobium* strain. Frontiers in Microbiology. 2017;**8**:2466. DOI: 10.3389/fmicb.2017.02466

[9] Dong R, Zhang J, Huan H, Bai C, Chen Z, Liu G. High salt tolerance of a *Bradyrhizobium* strain and its promotion of the growth of *Stylosanthes guianensis*. International Journal of Molelcular Sciences. 2017;**18**:1625. DOI: 10.3390/ijms18081625

[10] Ismail HM, Hussain A, Iqbal A, Khan SA, Lee IJ. Endophytic fungus *Aspergillus japonicus* mediates host plant growth under normal and heat stress conditions. BioMed Research International. 2018;**2018**:7696831. DOI: 10.1155/2018/7696831

[11] Mishra J, Prakash J, Arora N, Samuel O, Bernard O, Olubukola RG, et al. Role of beneficial soil microbes in sustainable agriculture and environmental management. World Journal of Microbiology and Biotechnology. 2016;**33**:1-16. DOI: 10.1007/s11274-017-2364-9

[12] Grover M, Ali SZ, Sandhya V, Rasul A, Venkateswarlu B. Role of microorganisms in adaptation of agriculture crops to abiotic stresses. World Journal of Microbiology and Biotechnology. 2010;**27**:1231-1240. DOI: 10.1007/s11274-010-0572-7

[13] Larson C. Losing arable land, China faces stark choice: Adapt or go hungry. Science. 2013;**339**:644-645. DOI: 10.1126/science.339.6120.644

[14] Coleman-Derr D, Tringe SG. Building the crops of tomorrow: Advantages of symbiont based approaches to improving abiotic stress tolerance. Frontiers in Microbiology. 2014;**5**:283. DOI: 10.3389/fmicb.2014.00283

[15] Zahran HH. *Rhizobium*-legume symbiosis and nitrogen fixation under severe conditions and in an arid climate. Microbiology and Molecular Biology Reviews. 1999;**63**:968-989

[16] Vincent B, Marlet S, Vidal A, Bouarfa S, Wu J, Yang J, et al. Water and soil salinity management and salt redistribution in irrigation systems. In: Proceedings 18th World Congress of Soil Science; 9-15 July; Philadelphia, Pennsylvania, USA; 2006

[17] Egamberdieva D, Gafurova L, Islam KR. Salinity effects on irrigated soil chemical and biological properties in the Syr Darya basin of Uzbekistan. In: Lal R, Sulaimanov M, Stewart B, Hansen D, Doraiswamy P, editors. Climate Change and Terrestrial C Sequestration in Central Asia. Taylor-Francis: New York; 2007. pp. 147-162. DOI: 10.1201/9780203932698.ch11

[18] Barnawal D, Bharti N, Maji D, Chanotiya CS, Kalra A. ACC deaminase-containing *Arthrobacter protophormiae* induces NaCl stress tolerance through reduced ACC oxidase activity and ethylene production resulting in improved nodulation and mycorrhization in *Pisum sativum*. Journal of Plant Physiology. 2014;**171**:884-894. DOI: 10.1016/j.jplph.2014.03.007

[19] Van Hoorn JW, Katerji N, Hamdy A, Mastororilli M. Effect of salinity on yield and nitrogen uptake of four grain legumes and on biological nitrogen contribution from the soil. Agricultural Water Management. 2001;**51**:87-98. DOI: 10.1016/S0378-3774(01)00114-7

[20] Hashem A, Abd Allah EF, Alqarawi AA, Al-Huqail AA, Wirth S, Egamberdieva D. The interaction between arbuscular mycorrhizal fungi and endophytic bacteria enhances plant growth of *Acacia gerrardii* under salt stress. Frontiers in Microbiology. 2016;7:1089. DOI: 10.3389/fmicb.2016.01089

[21] Manchanda G, Garg N. Salinity and its effects on the functional biology of legumes. Acta Physiologiae Plantarum. 2008;**30**:595-618. DOI: 10.1007/s11738-008-0173-3

[22] Waśkiewicz A, Muzolf-Panek M, Goliński P. Phenolic content changes in plants under salt stress. In: Ahmad P, Azooz MM, Prasad MNV, editors. Ecophysiology and Responses of Plants under Salt Stress. Springer: New York; 2013. pp. 283-314. DOI: 10.1007/978-1-4614-4747-4_11

[23] Subbarao GV, Johansen C, Jana MK, Rao DKK. Comparative salinity tolerance of symbiotically dependent at nitrogen fed pigeon pea (*Cajanus cajan*) and its wild relative *Atylosia platycarpa*. Biology and Fertility of Soils. 1990;**10**:11-16. DOI: 10.1007/BF00336118

[24] Zaidi A, Wani PA, Khan MS. Toxicity of heavy metals to legumes and bioremediation. Heidelberg: Springer; 2012

[25] Ahmad E, Zaidi A, Khan MS, Oves M. Heavy metal toxicity to symbiotic nitrogen-fixing microorganism and host legumes. In: Zaidi A, Wani PA, Khan MS, editors. Toxicity of Heavy Metals to Legumes and Bioremediation. Springer: Heidelberg; 2012. pp. 29-44. ISBN: 3709107296

[26] Hao X, Taghavi S, Xie P, Orbach MJ, Alwathnani HA, Rensing C, et al. Phytoremediation of heavy and transition metals aided by legume-rhizobia symbiosis. International Journal of Phytoremediation. 2014;**16**:179-202. DOI: 10.1080/15226514.2013.773273

[27] Mesa J, Mateos-Naranjo E, Caviedes MA, Redondo-Gómez S, Pajuelo E, Rodríguez-Llorente ID. Endophytic cultivable bacteria of the metal bioaccumulator *Spartina maritima*

improve plant growth but not metal uptake in polluted marshes soils. Frontiers in Microbiology. 2015b;**6**:1450. DOI: 10.3389/fmicb.2015.01450

[28] Mesa J, Rodríguez-Llorente ID, Pajuelo E, Barcia-Piedras JM, Caviedes MA, Redondo-Gómez S, et al. Moving closer towards restoration of contaminated estuaries: Bioaugmentation with autochthonous rhizobacteria improves metal rhizoaccumulation in native *Spartina maritima*. Journal of Hazardous Materials. 2015a;**300**:263-271. DOI: 10.1016/j.jhazmat.2015.07.006

[29] Navarro-Torre S, Barcia-Piedras JM, Mateos-Naranjo E, Redondo-Gómez S, Camacho M, Caviedes MA, et al. Assessing the role of endophytic bacteria in the halophyte *Arthrocnemum macrostachyum* salt tolerance. Plant Biology. 2017b;**19**:249-256. DOI: 10.1111/plb.12521

[30] Paredes-Páliz KI, Mateos-Naranjo E, Doukkali B, Caviedes MA, Redondo-Gómez S, Rodríguez-Llorente ID, et al. Modulation of *Spartina densiflora* plant growth and metal accumulation upon selective inoculation treatments: A comparison of gram negative and gram positive rhizobacteria. Marine Pollution Bulletin. 2017;**125**:77-85. DOI: 10.1016/j.marpolbul.2017.07.072

[31] Navarro-Torre S, Barcia-Piedras JM, Caviedes MA, Pajuelo E, Redondo-Gómez S, Rodríguez-Llorente ID, et al. Bioaugmentation with bacteria selected from the microbiome enhances *Arthrocnemum macrostachyum* metal accumulation and tolerance. Marine Pollution Bulletin. 2017a;**117**:340-347. DOI: 10.1016/j.marpolbul.2017.02.008

[32] Doyle JJ, Luckow MA. The rest of the iceberg. Legume diversity and evolution in a phylogenetic context. Plant Physiology. 2003;**131**:900-910. DOI: 10.1104/pp.102.018150

[33] FAO. Food and Agriculture Organization. Climate-Smart Agriculture Sourcebook. FAO. 2013. E-ISBN: 978-92-5-107721-4

[34] Pajuelo E, Rodríguez-Llorente ID, Lafuente A, Caviedes MÁ. Legume–*rhizobium* symbioses as a tool for bioremediation of heavy metal polluted soils. In: Khan MS, Zaidi A, Goel R, Musarrat J, editors. Biomanagement of Metal-Contaminated Soils. Springer: Heidelberg; 2011. pp. 95-123. ISBN: 978-94-007-1914-9

[35] Ghnaya T, Mnassri M, Ghabriche R, Wali M, Poschenrieder C, Lutts S, et al. Nodulation by *Sinorhizobium meliloti* originated from a mining soil alleviates Cd toxicity and increases Cd-phytoextraction in *Medicago sativa* L. Frontiers in Plant Science. 2015;**6**: 1-10. DOI: 10.3389/fpls.2015.00863

[36] Zribi K, Nouairi I, Slama I, Talbi-Zribi O, Mhadhbi H. *Medicago sativa—Sinorhizobium meliloti* symbiosis promotes the bioaccumulation of zinc in nodulated roots. International Journal of Phytoremediation. 2015;**17**:49-55. DOI: 10.1080/15226514.2013.828017

[37] Navarro-Torre S, Rodríguez-Llorente ID, Doukkali B, Caviedes MA, Pajuelo E. Competition for alfalfa nodulation under metal stress by the metal-tolerant strain *Ochrobactrum cytisi* Azn6.2. Annals of Applied Biology. 2019;**175**:184-192. DOI: 10.1111/aab.12528

[38] Dashti N, Khanafer M, Radwan SS. Endophytic and epiphytic hydrocarbon-utilizing bacteria associated with root nodules of legumes. In: Proceedings of the 28th Arctic and Marine Oilspill Program (AMOP) Technical Seminar; 7-9 July 2005. Calgary: Canada; 2005. pp. 1101-1109

[39] Dary M, Chamber-Pérez MA, Palomares AJ, Pajuelo E. "In situ" phytostabilisation of heavy metal

polluted soils using *Lupinus luteus* inoculated with metal resistant plant-growth promoting rhizobacteria. Journal of Hazardous Materials. 2010;**177**:323-330. DOI: 10.1016/j.jhazmat.2009.12.035

[40] Kong Z, Glick BR, Duan J, Ding S, Tian J, McConkey BJ, et al. Effects of 1-aminocyclopropane-1- carboxylate (ACC) deaminase-overproducing *Sinorhizobium meliloti* on plant growth and copper tolerance of *Medicago lupulina*. Plant and Soil. 2015;**70**:5891-5897. DOI: 10.1007/s11104-015-2434-4

[41] Raklami A, Oufdou K, Tahiri AI, Mateos-Naranjo E, Navarro-Torre S, Rodríguez-Llorente ID, et al. Safe cultivation of *Medicago sativa* in metal-polluted soils from semi-arid regions assisted by heat- and metallo-resistant PGPR. Microorganisms. 2019;**7**:212. DOI: 10.3390/microorganisms7070212

[42] Olanrewaju OS, Glick BR, Babalola OO. Mechanisms of action of plant growth promoting bacteria. World Journal of Microbiology and Biotechnology. 2017;**33**:197. DOI: 10.1007/s11274-017-2364-9

[43] Gouda S, Kerry RG, Das G, Paramithiotis S, Shin H, Patra JK. Revitalization of plant growth promoting rhizobacteria for sustainable development in agriculture. Microbiological Research. 2018;**206**:31-140. DOI: 10.1016/j.micres.2017.08.016

[44] Quiñones-Aguilar EE, Hernández Cuevas LV, López Pérez L, Rincón EG. Effectiveness of native arbuscular mycorrhizal fungi from Agave rhizosphere as growth promoters in papaya. TERRA Latinoamericana. 2019;**37**:163-174. DOI: 10.28940/tl.v37i2.397

[45] Cano MA. A review of interaction of beneficial microorganisms in plants: Mycorrhizae, *Trichoderma* spp. and *Pseudomonas* spp. Resvista UDCA Actualidad & Divulgación Científica. 2011;**14**:15-31

[46] Backer R, Rokem JS, Ilangumaran G, Lamont J, Praslickova D, Ricci E, et al. Plant growth-promoting rhizobacteria: Context, mechanisms of action, and roadmap to commercialization of biostimulants for sustainable agriculture. Frontiers in Plant Science. 2018;**9**:1-17. DOI: 10.3389/fpls.2018.01473

[47] Bashan Y, De-Bashan LE. Inoculant preparation and formulations for *Azospirillum* spp. In: Cassán FD, Okon Y, Creus CM, editors. Handbook for *Azospirillum*. Springer: Heidelberg; 2015. pp. 469-485. DOI: 10.1007/978-3-319-06542-7

[48] Santos AA, Silveira JAGD, Guilherme EA, Bonifacio A, Rodrigues AC, Figueiredo MDVB. Changes induced by co-inoculation in nitrogen-carbon metabolism in cowpea under salinity stress. Brazilian Journal of Microbiology. 2018;**49**:685-694. DOI: 10.1016/j.bjm.2018.01.007

[49] Saia S, Amato G, Frenda AS, Giambalvo D, Ruisi P. Influence of arbuscular mycorrhizae on biomass production and nitrogen fixation of berseem clover plants subjected to water stress. PLoS One. 2014;**9**:e90738. DOI: 10.1371/journal.pone.0090738

[50] Anad K, Kumari B, Mallick MA. Phosphate solubilizing microbes: An effective and alternative approach as. International Journal of Pharmacy and Pharmaceutical Sciences. 2016;**8**(2):37-40

[51] Alori ET, Glick BR, Babalola OO. Microbial phosphorus solubilization and its potential for use in sustainable agriculture. Frontiers in Microbiology. 2017;**8**:1-8. DOI: 10.3389/fmicb.2017.00971

[52] Zhu F, Qu L, Hong X, Sun X. Isolation and characterization of a phosphate- solubilizing halophilic bacterium *Kushneria* sp. YCWA18 from Daqiao saltern on the Coast of Yellow Sea of China. Evidence-based Complementary and Alternative Medicine. Marine Biotechhnology. 2011;**2011**:1-6. DOI: 10.1155/2011/615032

[53] Turan M, Kıtır N, Alkaya Ü, Günes A, Tüfenkçi Ş, Yıldırım E, et al. Making soil more accessible to plants: The case of plant growth promoting rhizobacteria. In: Rogovelo EC, editor. Plant Growth. IntechOpen: London; 2016. pp. 61-69. DOI: 10.5772/64826

[54] Anzuay MS, Ciancio MGR, Ludueña LM, Angelini JG, Barros G, Pastor N, et al. Growth promotion of peanut (*Arachis hypogaea* L.) and maize (*Zea mays* L.) plants by single and mixed cultures of efficient phosphate solubilizing bacteria that are tolerant to abiotic stress and pesticides. Microbiological Research. 2017;**199**:98-109. DOI: 10.1016/j.micres.2017.03.006

[55] Shahid M, Akram MS, Khan MA, Zubair M, Shah SM, Ismail M, et al. Phytobeneficial strain *Planomicrobium* sp. MSSA-10 triggered oxidative stress responsive mechanisms and regulated the growth of pea plants under induced saline environment. Journal of Applied Microbiology. 2018;**124**:1566-1579. DOI: 10.1111/jam.13732

[56] Vandana UK, Chopra A, Bhattacharjee S, Mazumder PB. Microbial biofertilizer: A potential tool for sustainable agriculture. In: Panpatte DG, Jhala YK, Vyas RV, Shelat HN, editors. Microorganisms for Green Revolution, Microorganisms for Sustainability. Vol. 6. Springer: Singapore; 2017. pp. 25-52. DOI: 10.1007/978-981-10-6241-4

[57] Egamberdieva D, Berg G, Lindström K, Räsänen LA. Alleviation of salt stress of symbiotic *Galega officinalis* L. (Goat's rue) by co-inoculation of *Rhizobium* with root colonising *Pseudomonas*. Plant and Soil. 2013;**369**:453-465. DOI: 10.1007/s11104-013-1586-3

[58] Wani PA, Khan MS, Zaidi A. Effect of metal tolerant plant growth promoting *Bradyrhizobium* sp. (vigna) on growth, symbiosis, seed yield and metal uptake by greengram plants. Chemosphere. 2007;**70**:36-45. DOI: 10.1016/j.chemosphere.2007.07.028

[59] Gupta G, Parihar SS, Ahirwar NK, Snehi SK, Singh V. Plant growth promoting rhizobacteria (PGPR): Current and future prospects for development of sustainable agriculture. Journal of Microbial & Biochemical Technology. 2015;7:96-102. DOI: 10.4172/1948-5948.1000188

[60] Selim SM, Zayed MS. Role of biofertilizers in sustainable agriculture under abiotic stresses. In: Panpatte DG, Jhala YK, Vyas RV, Shelat HN, editors. Microorganisms for Green Revolution, Microorganisms for Sustainability. Vol. 6. Springer: Singapore; 2017. pp. 281-301. DOI: 10.1007/978-981-10-6241-4_15

[61] Jha CK, Saraf M. Plant growth promoting Rhizobacteria (PGPR): A review. E3 Journal of Agricultural Research and Development. 2015;5:0108-0119. DOI: 10.13140/RG.2.1.5171.2164

[62] Benito P, Alonso-Vega P, Aguado C, Luján R, Anzai Y, Hirsch AM, et al. Monitoring the colonization and infection of legume nodules by *Micromonospora* in co-inoculation experiments with rhizobia. Science Reports. 2017;7:11051. DOI: 10.1038/s41598-017-11428-1

[63] Martínez-Hidalgo P, Hirsch A. The nodule microbiome: N2-fixing rhizobia do not live alone. Phytobiomes. 2017;**1**:70-82. DOI: 10.1094/PBIOMES-12-16-0019-RVW

[64] Kang BG, Kim WT, Yun HS, Chang SC. Use of plant growth-promoting rhizobacteria to control stress responses of plant roots. Plant Biotechnology Reports. 2010;**4**:179-183. DOI: 10.1007/s11816-010-0136-1

[65] Alori ET, Babalola OO. Microbial inoculants for improving crop quality and human health in Africa. Frontiers in Microbiology. 2018;**9**:1-12. DOI: 10.3389/fmicb.2018.02213

[66] Gurikar C, Naik MK, Sreenivasa MY. *Azotobacter*: PGPR activities with special reference to effect of pesticides and biodegradation. In: Singh DP, Singh HB, Prabha R, editors. Microbial Inoculants in Sustainable Agricultural Productivity. Springer: New Delhi; 2016. pp. 229-244. DOI: 10.1007/978-81-322-2647-5

[67] Saravanakumar D, Samiyappan R. ACC deaminase from *Pseudomonas fluorescens* mediated saline resistance in groundnut (*Arachis hypogea*) plants. Journal of Applied Microbiology. 2007;**102**:1283-1292. DOI: 10.1111/j.1365-2672.2006.03179.x

[68] Sarma RK, Saikia R. Alleviation of drought stress in mung bean by strain *Pseudomonas aeruginosa* GGRJ21. Plant and Soil. 2014;**377**:111-126. DOI: 10.1007/s11104-013-1981-9

[69] Barnawal D, Maji D, Bharti N, Chanotiya CS, Kalra A. ACC deaminase-containing *Bacillus subtilis* reduces stress ethylene-induced damage and improves mycorrhizal colonization and rhizobial nodulation in *Trigonella foenum-graecum* under drought stress. Journal of Plant Growth Regulation. 2013;**32**:809-822. DOI: 10.1007/s00344-013-9347-3

[70] Barnawal D, Bharti N, Maji D, Chanotiya CS, Kalra A. ACC deaminase-containing *Arthrobacter protophormiae* induces NaCl stress tolerance through reduced ACC oxidase activity and ethylene production resulting in improved nodulation and mycorrhization in *Pisum sativum*. Journal of Plant Physiology. 2014;**171**:884-894. DOI: 10.1016/j.jplph.2014.03.007

[71] Saikia J, Sarma RK, Dhandia R, Yadav A, Bharali R, Gupta VK, et al. Alleviation of drought stress in pulse crops with ACC deaminase producing rhizobacteria isolated from acidic soil of Northeast India. Scientific Reports. 2018;**8**:3560. DOI: 10.1038/s41598-018-21921-w

[72] Guo J, Chi J. Effect of Cd-tolerant plant growth-promoting *Rhizobium* on plant growth and Cd uptake by *Lolium multiflorum* lam. And *Glycine max* (L.) Merr. in Cd-contaminated soil. Plant and Soil. 2014;**375**:205-214. DOI: 10.1007/s11104-013-1952-1

[73] Patil S, Paradeshi J, Chaudhari B. Suppression of charcoal rot in soybean by moderately halotolerant *Pseudomonas aeruginosa* GS-33 under saline conditions. Journal of Basic Microbiology. 2016;**56**:889-899. DOI: 10.1002/jobm.201600008

[74] Kumar M, Mishra S, Dixit V, Kumar M, Agarwal L, Chauhan PS, et al. Synergistic effect of *Pseudomonas putida* and *Bacillus amyloliquefaciens* ameliorates drought stress in chickpea (*Cicer arietinum* L.). Plant Signaling & Behavior. 2016;**11**:e1071004. DOI: 10.1080/15592324.2015.1071004

[75] Sharma P, Khanna V, Kumari SK. Abiotic stress mitigation through plant-growth- promoting rhizobacteria. In: Choudhary D, Varma A, Tuteja N, editors. Plant-Microbe Interaction: An Approach to Sustainable Agriculture. Springer: Singapore; 2016. pp. 327-342. DOI: 10.1007/978-981-10-2854-0

[76] Hassen AI, Bopape FL, Sanger LK. Microbial inoculants as agents of growth promotion and abiotic stress tolerance in plants. In: Singh DP, Singh HB, Prabha R, editors. Microbial Inoculants

in Sustainable Agricultural Productivity. Springer: New Delhi; 2016. pp. 23-36. DOI: 10.1007/978-81-322-2647-5

[77] Karthik C, Elangovan N, Kumar TS, Govindharaju S, Barathi S, Oves M, et al. Characterization of multifarious plant growth promoting traits of rhizobacterial strain AR6 under chromium (VI) stress. Microbiological Research. 2017;**204**:65-71. DOI: 10.1016/j.micres.2017.07.008

[78] Martins SJ, Rocha GA, de Melo HC, Georg RC, Ulhôa CJ, Dianese EC, et al. Plant-associated bacteria mitigate drought stress in soybean. Environmetnal Science and Pollution Research International. 2018;**25**:13676-13686. DOI: 10.1007/s11356-018-1610-5

[79] Ulloa-Ogaz AL, Muñoz-Castellanos L, Nevárez-Moorillón GV. Biocontrol of phytopathogens: Antibiotic production as mechanism of control. In: Méndez-Vilas A, editor. The Battle against Microbial Pathogens: Basic Science, Technological Advances and Educational Programs. Vol. 1. Badajoz: Formatex; 2015. pp. 305-309

[80] Smitha M, Singh R. Biocontrol of phytopathogenic fungi using mycolytic enzymes produced by rhizospheric bacteria of *Cicer arietinum*. Indian Journal of Agricultural Biochemistry. 2014;**27**:215-218

[81] Liu Y, Baird SM, Qiao J, Du Y, Lu SE. SecG is required for antibiotic activities of *Pseudomonas* sp. YL23 against *Erwinia amylovora* and *Dickeya chrysanthemi*. Journal of Basic Microbiology. 2015;**55**(5):617-624. DOI: 10.1002/jobm.201400491

[82] Anwar MS, Paliwal A, Firdous N, Verma A, Kumar A, Pande V. Co-culture development and bioformulation efficacy of psychrotrophic PGPRs to promote growth and development of pea (*Pisum sativum*) plant. The Journal of General and Applied Microbiology. 2019;**65**: 88-95. DOI: 10.2323/jgam.2018.05.007

[83] Sabaté DC, Brandan CP, Petroselli G, Erra-Balsells R, Audisio MC. Biocontrol of *Sclerotinia sclerotiorum* (lib.) de Bary on common bean by native lipopeptide-producer *Bacillus* strains. Microbiological Research. 2018;**211**:21-30. DOI: 10.1016/j.micres.2018.04.003

[84] Goswami D, Dhandhukia P, Patel P, Thakker JN. Screening of PGPR from saline desert of Kutch: growth promotion in *Arachis hypogea* by *Bacillus licheniformis* A2. Microbiological Research. 2014;**169**(1):66-75. DOI: 10.1016/j.micres.2013.07.004

[85] Chaudhari BL, Patil SN, Paradeshi JS, Mangal A, Chaudhari CSC. Premier biocontrol traits of pseudomonads: Siderophores, phenazines or what else? In: Panpatte DG, Jhala YK, Vyas RV, Shelat HN, editors. Microorganisms for Green Revolution, Microorganisms for Sustainability. Vol. 6. Springer: Singapore; 2017. pp. 351-390. DOI: 10.1007/978-981-10-6241-4_18

[86] Fernandez-Göbel TF, Deanna R, Muñoz NB, Robert G, Asurmendi S, Lascano R. Redox systemic signaling and induced tolerance responses during soybean-*Bradyrhizobium japonicum* interaction: Involvement of nod factor receptor and autoregulation of nodulation. Frontiers in Plant Science. 2019;**10**:141. DOI: 10.3389/fpls.2019.00141

[87] Figueredo MS, Tonelli ML, Taurian T, Angelini J, Ibanez F, Valetti L, et al. Interrelationships between *Bacillus* sp. CHEP5 and *Bradyrhizobium* sp. SEMIA6144 in the induced systemic resistance against *Sclerotium rolfsii* and symbiosis on peanut plants. Journal of Biosciences. 2014;**39**:877-885. DOI: 10.1007/s12038-014-9470-8

[88] Pertot I, Puopolo G, Hosni T, Pedrotti L, Jourdan E, Ongena M. Limited impact of abiotic stress on surfactin production in planta and on disease

resistance induced by *Bacillus amyloliquefaciens* S499 in tomato and bean. FEMS Microbiology Ecology. 2013;**86**:505-519. DOI: 10.1111/1574-6941.12177

[89] Cornforth DM, Popat R, McNally L, Gurney J, Scott-Phillips TC, Ivens A, et al. Combinatorial *quorum sensing* allows bacteria to resolve their social and physical environment. PNAS. 2014;**111**:4280-4284. DOI: 10.1073/pnas.1319175111

[90] Sunar K, Dey P, Chakraborty U, Chakraborty B. Biocontrol efficacy and plant growth promoting activity of *Bacillus altitudinis* isolated from Darjeeling hills. India. Journal of Basic Microbiology. 2015;**55**:91-104. DOI: 10.1002/jobm.201300227

[91] Mohamed I, Eid KE, Abbas MHH, Salem AA, Ahmed N, Ali M, et al. Use of plant growth promoting rhizobacteria (PGPR) and mycorrhizae to improve the growth and nutrient utilization of common bean in a soil infected with white rot fungi. Ecotoxicology and Environmental Safety. 2019;**171**:539-548. DOI: 10.1016/j.ecoenv.2018.12.100

[92] Huang J, Shi Y, Zeng G, Gu Y, Chen G, Shi L, et al. Acyl-homoserine lactone-based *quorum sensing* and *quorum quenching* hold promise to determine the performance of biological wastewater treatments: An overview. Chemosphere. 2016;**157**:137-151. DOI: 10.1016/j.chemosphere.2016.05.032

[93] Fetzner S. *Quorum quenching* enzymes. Journal of Biotechnoly. 2015;**201**:2-14. DOI: 10.1016/j.jbiotec.2014.09.001

[94] Hartmann A, Rothballer M, Hense BA, Schröder P. Bacterial quorum sensing compounds are important modulators of microbe-plant interactions. Frontiers in Plant Science. 2014;**5**:1-4. DOI: 10.3389/fpls.2014.00131

[95] Tiwari S, Shweta S, Prasad M, Lata C. Genome-wide investigation of GRAM-domain containing genes in rice reveals their role in plant-rhizobacteria interactions and abiotic stress responses. International Journal of Biological Macromolecules. 2019. DOI: 10.1016/j.ijbiomac.2019.11.162 (in press)

[96] Manoj SR, Karthik C, Kadirvelu K, Arulselvi PI, Shanmugasundaram T, Bruno B, et al. Understanding the molecular mechanisms for the enhanced phytoremediation of heavy metals through plant growth promoting rhizobacteria: A review. Journal of Environmental Management. 2020;**254**:109779. DOI: 10.1016/j.jenvman.2019.109779

[97] Ghassemi HR, Mostajeran A. *TASOS1* and *TATM20* genes expression and nutrient uptake in wheat seedlings may be altered via excess cadmium exposure and inoculation with *Azospirillum brasilense* sp. 7 under saline condition. Applied Ecology and Environmental Research. 2018;**16**:1797-1817. DOI: 10.15666/aeer/1602_17971817

[98] Ambreetha S, Chinnadurai C, Marimuthu P, Balachandar D. Plant-associated Bacillus modulates the expression of auxin-responsive genes of rice and modifies the root architecture. Rhizosphere. 2018;**5**:57-66. DOI: 10.1016/j.rhisph.2017.12.001

[99] Lafuente A, Pérez-Palacios P, Doukkali B, Molina-Sánchez MD, Jiménez-Zurdo JI, Caviedes MA, et al. Unraveling the effect of arsenic on the model Medicago–Ensifer interaction: A transcriptomic meta-analysis. The New Phytologist. 2015;**205**:255-272. DOI: 10.1111/nph.13009

[100] Lafuente A, Pajuelo E, Caviedes MA, Rodríguez-Llorente ID. Reduced nodulation in alfalfa induced by arsenic correlates with altered expression of early nodulins. Journal

of Plant Physiology. 2010;**167**:286-291. DOI: 10.1016/j.jplph.2009.09.014

[101] Khanna K, Jamwal VL, Kohli SK, Gandhi SG, Ohri P, Bhardwaj R, et al. Plant growth promoting rhizobacteria induced Cd tolerance in *Lycopersicon esculentum* through altered antioxidative defense expression. Chemosphere. 2019;**217**:463-474. DOI: 10.1016/j.chemosphere.2018.11.005

[102] Govindasamy V, George P, Kumar M, Aher L, Raina SK, Rane J, et al. Multi-trait PGP rhizobacterial endophytes alleviate drought stress in a senescent genotype of sorghum [*Sorghum bicolor* (L.) Moench]. 3 Biotech. 2020;**10**:13. DOI: 10.1007/s13205-019-2001-4

[103] Khan N, Bano A, Rahman MA, Guo J, Kang Z, Bbar MA. Comparative physiological and metabolic analysis reveals a complex mechanism involved in drought tolerance in chickpea (*Cicer arietinum* L.) induced by PGPR and PGRs. Scientific Reports. 2019;**9**:2097. DOI: 10.1038/s41598-019-38702-8

[104] Chauhan PS, Lata C, Tiwari S, Chauchan AS, Mishra K, Agrawal L, et al. Transcriptional alterations reveal *Bacillus amyloliquefaciens*-rice cooperation under salt stress. Scientifics Report. 2019;**9**:11912. DOI: 10.1038/s41598-019-48309-8

[105] Bharti N, Pandey S, Barnawal D, Patel VK, Kalra A. Plant growth promoting rhizobacteria *Dietzia natronolimnaea* modulates the expression of stress responsive genes providing protection of wheat from salinity stress. Scientific Reports. 2016;**6**:34768. DOI: 10.1038/srep34768

[106] Gao J, Luo M, Peng H, Chen F, Li W. Characterization of cadmium-responsive MicroRNAs and their target genes in maize (*Zea mays*) roots. BMC Molecular Biology. 2019;**20**:14. DOI: 10.1186/s12867-019-0131-1

[107] Wang B, Cheng D, Chen Z, Zhang M, Zhang G, Jiang M, et al. Bioinformatic exploration of the targets of xylem sap miRNAs in maize under cadmium stress. International Journal of Molecular Sciences. 2019;**20**:1474. DOI: 10.3390/ijms20061474

[108] Jatan R, Tiwari S, Asif MH, Lata C. Genome-wide profiling reveals extensive alterations in *Pseudomonas putida* mediated miRNAs expression during drought stress in chickpea (*Cicer arietinum* L.). Environmental and Experimental Botany. 2019;**157**:217-227. DOI: 10.1016/j.envexpbot.2018.10.003

Chapter 8

Bioactive Components of Magical Velvet Beans

Suresh S. Suryawanshi, Prajakta P. Kamble, Vishwas A. Bapat and Jyoti P. Jadhav

Abstract

The plant *Mucuna* is an annual climbing shrub with long vines that can reach over fifteen meters in length. About 100–150 *Mucuna* species are found in the tropic and subtropic regions of both hemispheres of the earth. The genus *Mucuna* belongs to the family Leguminosae. It is commonly known as Kewanch, velvet bean, cowhage and kappikachhu and is found widely in India as a hardy, herbaceous, vigorous, twining annual plant. The size and dimension of the *Mucuna* seeds, pods, platelets and leaves change from species to species. The hair present on pods is anthelmintic, which causes itching. People are seeking great attention towards *Mucuna* due to its several medicinal properties, including L-DOPA (L-3, 4-dihydroxyphenylalanine) along with supplementary antioxidants that are used for treating Parkinson's disease and many neurodegenerative diseases. Thus it is being used in about 200 medicinal formulations. The current chapter outlines the work that determines the influence of different nutritional, anti-nutritional and medicinal values and bioactive agents from different parts of the *Mucuna* species present in India and its importance in medicine.

Keywords: Legumes, *Mucuna*, Parkinson's disease, L-dopa, antioxidants

1. Introduction

Mucuna is a valuable genus of Leguminosae family; it has tremendous value in food and medicine. The genus *Mucuna* originated in Eastern India and China and then was transferred throughout tropical and subtropical regions of the world creating new populations [1]. Traditionally, *Mucuna* is known by different names including beans, buffalo beans, dopa bean, cowitch, kappikachhu and atmagupta. Out of 100 *Mucuna* species found worldwide, 8 species and 3 varieties of *Mucuna* are predominantly found in different localities of India [1–4]. The *Mucuna* plant is an annual perennial climbing shrub with long vines having a length of 15–25 m with trifoliate leaves and yielding long inflorescences with purple or pale yellow flowers. They produce green or brown pods covered with rigid hair, which causes intense itching [5]. Pods contain four to six ellipsoidal-shaped seeds that are rich dark brown or blotched [6] varying from species to species. *Mucuna* seeds are a rich source of nutritional, antinutritional and phytochemical compounds containing L-dopa as a prime constituent [7]. The content of L-dopa varies between from species to species and locality to locality. Among them, *M. pruriens* is the most exploited species as a remedy against Parkinson's disease [8]. Due to huge international and national trade price and scarcity of *M. pruriens*, other *Mucuna* species are reported to be adulterants for *M. pruriens.*

Ancient reports of Ayurveda suggest that *Mucuna* seed powder contributes in reducing the risks of certain cardiovascular diseases and neurodegenerative disease and also as a remedy for snake bite. The seeds of *Mucuna* have gained increasing attention among food scientists, nutrition specialists and pharmaceutical expertise due to their rich source of antioxidant, phenolic, flavonoids, L-dopa, proteins, starch, micronutrients, dietary fiber and bioactive compounds that play a pronounced role in the traditional as well as modern medicine all over the world [9, 10]. The existing scenario shows ten reported *Mucuna* species that were studied recently by Pulikkalpura et al., from the Indian subcontinent [11], whereas Patil et al., also collected fourteen different species of *Mucuna* from various localities and further studied for their L-dopa content (anti-Parkinson's activity) [6, 12].

Mucuna grows best under moist, warm conditions and in areas with plenteous rainfall. It can propagate in any type of soil but sandy lome soil is mostly favorable with pH of 5.5–7.5. Several researchers have investigated different species having typical characteristics like size and shape of bracts, leaflets and pods, color, thickness, density as well as number of seeds in pod and flower, respectively [6]. The evaluation of genetic-level studies of *Mucuna* species from India was also carried out using inter-simple sequence repeat markers and randomly amplified polymorphic DNA [13]. The seeds show tolerance against different abiotic stress including low soil fertility and acidic pH and also grow in wet soils (Duke, 1981). Similar to different species, *Mucuna* also has 2n = 2x = 22 number of chromosomes with genome size ranging between 1281 and 1361 Mbp/C [14]. Due to wild fluctuating climatic and geographical distribution, these species show gigantic diversity in phenotype in the Indian subcontinent. Corresponding to the family *Leguminosae*, it also has the ability of atmospheric nitrogen fixation. It is also grown for the potential utility in animal feed and human food due to its rich source of nutritional content [15, 16]. Thus, it was consumed universally for the treatment of Parkinson's disease. Traditionally, in Ayurvedic science, *Mucuna* (velvet bean) plant is widely used to treat numerous diseases including parkinsonism [17–21] due to its L-dopa content as one of the principal constituents [9]. All parts of *Mucuna* have a great medicinal value in the ancient traditional medicinal system, and hence, it has a prodigious demand in the international and the Indian market [7, 22].

Mucuna is a superb source of protein and bioactive compounds that have increased consumption per capita after being considered as a functional food by the US [23]. The previous literature survey shows that the declining occurrence of numerous long-lasting disorders, namely neurological disorder, cardiovascular diseases, diabetes, obesity and cancer, has a positive correlation with the consumption of legume seeds [24]. Considering all the evident health profits, studying its bioactive compounds is of great importance. Among all the under-utilized *Mucuna* species, normally available and commonly used *Mucuna pruriens* seeds were studied enormously and have been reported in numerous of articles published till date. To avoid the burden on commonly available and used *Mucuna Purience,* various researchers are studying bioactive components and the use of other *Mucuna* species like *M. imbracata*, *M. bracteata*, *M. monosperma*, *M. macrocarpa, M. sanjappae,* M. atropurpurea, M. nigricans, M. gigantea, M. pruriens var *hirsuta, M. laticifera, M. yadaviana*, etc. in the treatment of various diseases [5–7, 9, 11–14, 17, 20, 22, 25–34]. Phytochemistry, toxilogicalogycal and food potential on the *Mucuna* species under study in the world were described by Lorenzetti et al. [35].

2. Bioactive compounds from various parts of *Mucuna* species

2.1 Bioactive compounds from seeds of *Mucuna*

Seeds of *Mucuna* are commonly used part of the plant, which are a rich source of nutritional and anti-nutritional compounds like L-dopa (anti-Parkinson's activity), antioxidants, phenolic, flavonoids, tannin, carbohydrates, starch, protein, micronutrients Sopanines and many more [9, 12, 26]. Antioxidant activities in this plant are mainly due to phenolic and various bioactive compounds present in the seed material [36, 37]. There are various extraction techniques, different solvents and processing methods that were used to extract the biologically active compounds from the seed of *Mucuna* [26, 31]. The prior study reports that *M. macrocarpa, M. sanjappae* and *M. atropurpurea* disclosed a higher level of L-dopa content, which also concludes that L-dopa content has a positive correlation to the protein content of seeds [12]. These high-yielding varieties of *Mucuna* can be commercially cultivated, which can thus serve to be a good option to lower the burden exerted on commonly used *M. prurience* variety [28]. LCMS analysis of four different species demonstrates the presence of diverse group of phenolics, alkaloids, flavonoids, different derivatives of gallic acid, L-dopa, catechin, alkaloids, quercetin, tannic acid, glycosides, saponins, tubastatin and a variety of amino acids in the seed extract [38]. Apart from that, it is also concluded that few anticancer compounds like *Spergualin,* sanggenon G, isopentenyl adenosine and spisulosine are also present in the seed extract [28, 38].

2.2 Bioactive compounds from leaves and roots of *Mucuna*

The root extract of *Mucuna* has various activities like stimulant, thermogenic, purgative, emollient, diuretic anthelmintic, emmenagogue and tonic; hence, they are used in the vitiated circumstances in Veda and Ayurveda [39]. L-Dopa content of leaves and roots is as much as 1% and 4–7% in *Mucuna* plant [40]. *Mucuna* plants release secondary chemical compounds called allelochemical in the form of L-3,4-dihydroxyphenylalanine (L-dopa) in the surrounding environment which show an impact on growth of nearby plants, either negatively or positively. These substances are produced through its roots, seeds or leaves [41]. These secondary chemical agents play a role in damaging root growth, terminating seedling growth, inhibiting plantlet growth or suppressing seed germination of other plants [42]. Plant-box bioassay explains that the secondary chemical compound produced from the root of *Mucuna* is L-dopa [40], which affects the cell and root of various plant seedlings [41]. Leaf extracts are used to treat various complications like Anticataleptic, antiepileptic, aphrodisiac, antimicrobial, tonic and ulcers are some applications in which *Mucuna* leaves were reported being used previously [8, 20, 43, 44].

2.3 Bioactive compounds from callus of *Mucuna*

Production of callus from *Mucuna* plant material is a new era in the advancement of biochemical engineering and industrial biotechnology, which has the potential to produce different biologically active agents from the explant [45, 46]. Their application in cost-effective industrially important product formation is helpful for humankind, which upshots effective drug formulations and upsurges the nutritional level of food [47, 48]. L-Dopa is a major component in different parts of *Mucuna* species [9]. This also helps in storing germplasm of endangered species, which in turn leads to regenerate new plantlets at any time. Production of callus from *Mucuna* species was done previously by the researcher at the lab scale [49]. An earlier study by Chattopadhyay et al. depicted the formation of callus culture of commonly used

Mucuna pruriens [50]. Media containing various concentrations of media components affects biologically active chemicals and growth of the callus [51, 52]. The use of different elicitors and precursors are studied by Nandeo and Patel et al. [53, 54]. Implementation of precursor in the media of callus enhances the phenolic content as reported in prior studies. The percentage of phenolic is greater in callus culture than in seeds, which is very helpful for industrial production [55, 56].

2.4 Bioactive compounds from cell suspension of *Mucuna*

Reactive oxygen species (ROS) and reactive nitrogen species (RNS) are stress-producing free radicals, but at usual level perform an important part in the physiology of the body [57] to accommodate the massive demands for L-dopa and other secondary metabolites. *In vitro* production of biologically active compounds from suspension culture is predominantly studied before in *Mucuna* plant by Chattopadhyay et al. [50]. The use of mevalonic acid and its precursor gibberellic acid affect the growth. L-Dopa in callus exhibits a declining trend in fresh weight with a rise in concentration of L-dopa as shown by Desai et al. [52]. The comparative study of production of L-dopa from cell suspension culture and effect of elicitor on two different species like *Mucuna pruriens* L. and *Mucuna prurita* H were also done previously [58]. Large-scale production of phytochemicals and L-dopa was done from *Mucuna pruriens* L. Commercial production of the drugs (L-dopa), phenolic flavonoids and antioxidants using cell suspension cultures is in extensive practice nowadays.

3. Use of different bioactive compounds from *Mucuna* in various application

3.1 Antioxidant activity of *Mucuna* species

Numerous studies on antioxidant activity and phytoconstituents content of *Mucuna* seeds, leaves and roots were performed previously [4, 7, 19, 30, 31, 59–62]. Optimization of different solvents for the extraction of antioxidants was done by Aware and Patil et al. [9, 10, 26]. They concluded that depending on solubility of antioxidant compounds present in different species of *Mucuna*, concentration of antioxidants differs. Most commonly, ethanolic extract of *Mucuna* shows good antioxidant activity due to high phenolic content [43]. Some reports also conclude that water is a universal solvent, which shows significant quantity of phenolic, flavonoids and strong antioxidants that can scavenge free radicals using different assays. Diseases like neurodegenerative diseases, cardiovascular diseases, aging, cancer, rheumatoid arthritis and inflammation are caused by oxidative stresses, which are protected by ROS and RNS [60]. LCMS report of four different species of *Mucuna* determines the presence of various components like phenolic, flavonoids and bioactive compounds, which are responsible for production of reactive species [38].

3.2 Antimicrobial activity of *Mucuna* species

There are several compounds in *Mucuna* that contribute for the antimicrobial activity as reported in a previous study [38, 43]. These compounds are responsible for the treatment of various infectious diseases and ulcers [63]. The study on various plant pathogens suggests that methanolic extract of *Mucuna pruriens* seeds showed highest antimicrobial activity [64] from all used solvents. A similar type of study was done by Pujari *et al.*, who concluded that methanol extract of seeds of *Mucuna pruriens* was found to impart the best inhibiting activity among all scrutinized pathogens as compared to ethanol and acetone solvents. But alcoholic extract of *Mucuna pruriens* (L.) leaves has significant antioxidant and antibacterial activity, which has strongly recommended the use of *Mucuna* leaves and seed extract in traditional as well as modern medicine [65].

3.3 Protective effect against snake venom

Snakebite kills countless people annually since ancient days [66]. Various reports show the cross-reactivity between the enzyme of snake venom and protein from *Mucuna*, which determines the activity of *Mucuna* against snake venom [32, 67–69]. Betancur *et al.* in their review on therapeutics of antisnake venom explain the effectiveness of herbal plants, which act as coadjuvants and thus help to nullify the venom toxic action [68]. In recent literature, Kasturiratne *et al.* studied the global scenario of snakebite and deaths. They also elucidate that various traditional medicines were sometimes preferred with western drugs [70]. The protective effect of *Mucuna* in a study on mice or rat models proves that it has a good activity for curing snake bite, than few reported antivenom [71, 72].

3.4 Anti-Parkinson's activity of *Mucuna* species

Parkinson's disease (PD) was initially discovered by Dr. James Parkinson in 1817. It is a chronic neurological disorder triggered by a progressive loss of dopaminergic neurons present in the nigrostriatal part of the brain and found to be common in the US [73]. The major signs of the disease are complications in body movements, speaking, walking and many more complications arise as the disease progresses. Anti-Parkinson's potential of *Mucuna* is well known from ancient times due to its L-dopa content [7, 74]. L-Dopa is a precursor of dopamine used in the treatment of neurodegenerative disorders. Various scientists studied the potential of *Mucuna* to produce L-dopa as a source of anti-Parkinson's drug [8, 19, 75]. L-Dopa with other phytochemical compounds has a cumulative effect on the management of Parkinson's disease. Patil *et al.* describe that there is a correlation among the L-dopa, protein and carbohydrate content [12]. *Mucuna* is a rich source of antioxidant compounds, which performs a very important role in the physiology of the body mainly functioning in the inhibition of damage occurred because of free radicals [76]. There are hundreds of compounds that function as antioxidants in the plant system mainly vitamins, polyphenols, enzymes, flavonoids and metals like zinc, selenium, etc. [77]. The efficiency of the use of L-dopa and another dopaminergic agent in the treatment of Parkinson's disease is reviewed previously by Koller and Rueda [78]. The use of plants for the treatment is more beneficial than chemically manufactured medicines due to their infinitesimal occurrence of secondary complications by routine use and economical feasibility.

3.5 Use of *Mucuna* species in soil fertility

Cover crops have a role in the nitrogen-fixing bacteria and improvement of soil fertility by restoration of soil nutrients. Enormous use of chemical fertilizer and water in soil makes soil infertile, to overcome this problem, farmers are implementing traditional methods to enhance soil fertility. Mucuna is one of the best examples of a cover crop that has a rich source of biological natural products, which will increase the enhance soil fertility and fix atmospheric nitrogen [35].

4. Conclusion

Mucuna is a medicinally and biochemically valuable plant used from ancient days, having a large market value due to the presence of a large number of bioactive compounds. The content of phytochemical compounds and other bioactive agents present in *Mucuna* fluctuates from species to species. L-Dopa is a chiefly present amino acid found abundantly in *Mucuna* plant, which is used for the treatment of Parkinson's disease. It also contains a great amount of phenolics, flavonoids and antioxidants, which play a role in releasing oxidative stress. It also acts as a protein-rich diet. Due to all these properties, *Mucuna* has several applications in the pharmaceutical and food industries thereby uplifting the demand of *Mucuna* in day to day life.

Acknowledgements

Suresh Suryawanshi sincerely thanks the Indian Council of Medical Research (ICMR) for providing SRF fellowship [Award No. 45/6/2019/MP/BMS]. Vishwas A. Bapat thanks the National Academy of Sciences, India, for the honorary scientist fellowship. Prof. Jyoti P. Jadhav sincerely thanks the Department of Biotechnology, Government of India, for "Interdisciplinary Programme of Life Sciences for Advanced Research and Education (IPLS-Reference No: BT/PR4572/INF/22/147/2012)".

Conflict of interest

The authors declare no conflict of interest.

Author details

Suresh S. Suryawanshi[1], Prajakta P. Kamble[2], Vishwas A. Bapat[3] and Jyoti P. Jadhav[1,3]*

1 Department of Biochemistry, Shivaji University, Kolhapur, Maharashtra, India

2 Department of Microbiology, Shivaji University, Kolhapur, Maharashtra, India

3 Department of Biotechnology, Shivaji University, Kolhapur, Maharashtra, India

*Address all correspondence to: biochemjpj@gmail.com

References

[1] Wilmot-Dear CM. A revision of Mucuna (Leguminosae-Phaseoleae) in China and Japan. Kew Bulletin. 1984;**39**:23-25. DOI: 10.2307/4107853

[2] Wilmot-Dear CM. A revision of Mucuna (Leguminosae: Phaseoleae) in Thailand, Indochina and the Malay Peninsula. Kew Bulletin. 1992;**47**:203-245. DOI: 10.2307/4110664

[3] Ren S, Wilmot-dear CM, Browne SP, Browne ZP. About 100 Species: Worldwide; 18 Species (Nine Endemic) in China, Including Two Incompletely Known Taxa and One Possibly Extinct Species (*Mucuna* championii). 2010. pp. 207-218

[4] Aitawade MM, Yadav SR. Mucuna sanjappae, a new species from the north-Western Ghats, India. Kew Bulletin. 2012;**67**:539-543. DOI: 10.1007/s12225-012-9369-1

[5] Leelambika M, Sathyanarayana N. Genetic characterization of indian mucuna (Leguminoceae) species using morphometric and random amplification of polymorphic DNA (RAPD) approaches. Plant Biosystems. 2011;**145**:786-797. DOI: 10.1080/11263504.2011.610924

[6] Gaikwad SV, Gurav RV, Yadav SR. Karyotype studies in *Mucuna macrocarpa* Wall and *Mucuna sanjappae* Aitawade et Yadav (Fabaceae) from India. Chromosome Botany. 2017;**12**:52-55. DOI: 10.3199/iscb.12.52

[7] Kavitha C, Thangamani C. Amazing bean Mucuna pruriens: A comprehensive review. Journal of Medicinal Plant Research. 2014;**8**:138-143. DOI: 10.5897/jmpr2013.5036

[8] Lampariello L, Cortelazzo A, Guerranti R, Sticozzi C, Valacchi G. The magic velvet bean of mucuna pruriens. Journal of Traditional and Complementary Medicine. 2012;**2**:331-339. DOI: 10.1016/S2225-4110(16)30119-5

[9] Rane M, Suryawanshi S, Patil R, Aware C, Jadhav R, Gaikwad S, et al. Exploring the proximate composition, antioxidant, anti-Parkinson's and anti-inflammatory potential of two neglected and underutilized Mucuna species from India. South African Journal of Botany. 2019;**124**:304-310. DOI: 10.1016/j.sajb.2019.04.030

[10] Patil RR, Rane MR, Bapat VA, Jadhav JP. Phytochemical analysis and antioxidant activity of *Mucuna sanjappae*: A possible implementation in the parkinson' s disease treatment. Journal of Pharmaceutical and Medicinal Research. 2016;**2**:48-51

[11] Pulikkalpura H, Kurup R, Mathew PJ, Baby S. Levodopa in Mucuna pruriens and its degradation. Scientific Reports. 2015;**5**:2-10. DOI: 10.1038/srep11078

[12] Patil R, Aware C, Gaikwad S, Rajebhosale M, Bapat V, Yadav S, et al. RP-HPLC analysis of anti-Parkinson's drug l-DOPA content in Mucuna species from Indian subcontinent. Proceedings of the National Academy of Sciences, India Section B: Biological Sciences. 2019. DOI: 10.1007/s40011-018-01071-9

[13] Patil RR, Pawar KD, Rane MR, Yadav SR, Bapat VA, Jadhav JP. Assessment of genetic diversity in Mucuna species of India using randomly amplified polymorphic DNA and inter simple sequence repeat markers. Physiology and Molecular Biology of Plants. 2016;**22**:207-217. DOI: 10.1007/s12298-016-0361-3

[14] Bairiganjan GC, Patnaik SN. Chromosomal evolution in Fabaceae. Cytologia (Tokyo). 1989;**54**:51-64. DOI: 10.1508/cytologia.54.51

[15] Tuleun CD, Patrick JP, Tiamiyu LO. Evaluation of raw and boiled velvet bean (*Mucuna utilis*) as feed ingredient for broiler chickens. Pakistan Journal of Nutrition. 2009;**8**:601-606. DOI: 10.3923/pjn.2009.601.606

[16] Harms RH, Simpson CF, Waldroup PW. Influence of feeding various levels of velvet beans to chicks and laying hens. The Journal of Nutrition. 1961;**75**:127-131. DOI: 10.1093/jn/75.1.127

[17] Sathiyanarayanan L, Arulmozhi S. Mucuna pruriens Linn.—A comprehensive review. Pharmacognosy Reviews. 2007;**1**:157-162

[18] Suresh S, Prakash S. Effect of *Mucuna pruriens* (Linn.) on sexual behavior and sperm parameters in streptozotocin-induced diabetic male rat. Journal of Sexual Medicine. 2012;**9**:3066-3078. DOI: 10.1111/j.1743-6109.2010.01831.x

[19] Yadav K, Mukesh PU, Purohit S, Pandey B, Shah H. Phytochemistry and pharmacological activity of *Mucuna pruriens*: A review. International Journal of Green Pharmacy. 2017;**11**:69-723. DOI: 10.22377/ijgp.v11i02.916

[20] Torabi M, Drahansky M, Paridah M, Moradbak A, Mohamed A, Owolabi FAT, et al. We are IntechOpen, the world' s leading publisher of Open Access books Built by scientists, for scientists TOP 1 %, Intech. 2016;**i**:13. DOI: 10.5772/57353

[21] Contin M, Lopane G, Passini A, Poli F, Iannello C, Guarino M. Mucuna pruriens in Parkinson Disease. Clinical Neuropharmacology. 2015;**38**:201-203. DOI: 10.1097/WNF.0000000000000098

[22] Buckles D. Velvetbean: A "New" Plant with a History Author(s): Daniel Buckles Published by: Springer on behalf of New York Botanical Garden Press Stable. Preserve and extend access to Economic Botany All use subject t. 1995;**49**:13-25. Available from: https://www.jstor.org/stable/4255685

[23] Luthria DL, Pastor-Corrales MA. Phenolic acids content of fifteen dry edible bean (*Phaseolus vulgaris* L.) varieties. Journal of Food Composition and Analysis. 2006;**19**:205-211. DOI: 10.1016/j.jfca.2005.09.003

[24] Troszyńska A, Estrella I, López-Amóres ML, Hernández T. Antioxidant activity of pea (*Pisum sativum* L.) seed coat acetone extract. LWT—Food Science and Technology. 2002;**35**:158-164. DOI: 10.1006/fstl.2001.0831

[25] Patil RR, Gholave AR, Jadhav JP, Yadav SR, Bapat VA. Mucuna sanjappae Aitawade et Yadav: A new species of Mucuna with promising yield of anti-Parkinson's drug L-DOPA. Genetic Resources and Crop Evolution. 2014;**62**:155-162. DOI: 10.1007/s10722-014-0164-8

[26] Aware C, Patil R, Gaikwad S, Yadav S, Bapat V, Jadhav J. Evaluation of L-dopa, proximate composition with in vitro anti-inflammatory and antioxidant activity of Mucuna macrocarpa beans: A future drug for Parkinson treatment. Asian Pacific Journal of Tropical Biomedicine. 2017;**7**:1097-1106. DOI: 10.1016/j.apjtb.2017.10.012

[27] Patil SA, Surwase SN, Jadhav SB, Jadhav JP. Optimization of medium using response surface methodology for l-DOPA production by Pseudomonas sp. SSA. Biochemical Engineering Journal. 2013;**74**:36-45. DOI: 10.1016/j.bej.2013.02.021

[28] Raina AP, Misra RC. Chemical evaluation of Mucuna species for L-dopa content—An anti-Parkinson's drug yielding medicinal plant from India. Indian Journal of Traditional Knowledge. 2018;**17**:148-154

[29] Sathyanarayana N, Pittala RK, Tripathi PK, Chopra R, Singh HR, Belamkar V, et al. Transcriptomic resources for the medicinal legume *Mucuna pruriens*: De novo transcriptome assembly, annotation, identification and validation of EST-SSR markers. BMC Genomics. 2017;**18**:1-18. DOI: 10.1186/s12864-017-3780-9

[30] Misra L, Wagner H. Extraction of bioactive principles from *Mucuna pruriens* seeds. Indian Journal of Biochemistry & Biophysics. 2007;**44**:56-60

[31] Vadivel V, Biesalski HK. Bioactive compounds in velvet bean seeds: Effect of certain indigenous processing methods. International Journal of Food Properties. 2012;**15**:1069-1085. DOI: 10.1080/10942912.2010.513466

[32] Pavan H. A new host record for Uromyces mucunae on Mucuna sanjappae. Plant Pathology and Quarantine. 2014;**4**:90-91. DOI: 10.5943/ppq/4/2/2

[33] Phylogeny L, Group W, Jeff J, Toby R, Michael J, Martin F, et al. Legume phylogeny and classification in the 21st century: Progress, prospects and lessons for other species-rich clades. Taxon. 2013;**62**:217-248. DOI: 10.12705/622.8

[34] Cilia R, Laguna J, Cassani E, Cereda E, Pozzi NG, Isaias IU, et al. Mucuna pruriens in Parkinson disease: A double-blind, randomized, controlled, crossover study. Neurology. 2017;**89**:432-438. DOI: 10.1212/WNL.0000000000004175

[35] Lorenzetti F, Macisaac S, Arnason JT, Buckles D. The phytochemistry, toxicology, and food potential of velvetbean (Mucuna Adans. spp., Fabaceae). Africa (Lond). 2010:1-11

[36] Biradar SP, Tamboli AS, Khandare RV, Pawar PK. Chebulinic acid and Boeravinone B act as anti-aging and anti-apoptosis phyto-molecules during oxidative stress. Mitochondrion. 2019;**46**:236-246. DOI: 10.1016/j.mito.2018.07.003

[37] Thiruvengadam M, Rekha K, Rajakumar G, Lee T-J, Kim S-H, Chung I-M. Enhanced production of anthraquinones and phenolic compounds and biological activities in the cell suspension cultures of *Polygonum multiflorum*. International Journal of Molecular Sciences. 2016;**17**:1912. DOI: 10.3390/ijms17111912

[38] Suryawanshi JPJSS, Rane MR, Kshirsagar PR, Kamble PP. Antioxidant, antimicrobial activity with mineral composition and LCMS based phytochemical evaluation of some Mucuna species from India. International Journal of Pharmacy and Biological Sciences. 2019;**9**:312-324

[39] Kumar S, Dobos GJ, Rampp T. The significance of ayurvedic medicinal plants. Evidence-Based Complementary and Alternative Medicine. 2017;**22**:494-501. DOI: 10.1177/2156587216671392

[40] Nishihara E, Parvez MM, Araya H, Kawashima S, Fujii Y. L-3-(3,4-Dihydroxyphenyl)alanine (L-DOPA), an allelochemical exuded from velvetbean (*Mucuna pruriens*) roots. Plant Growth Regulation. 2005;**45**:113-120. DOI: 10.1007/s10725-005-0610-x

[41] Guidotti BB, Gomes BR, Siqueira-Soares RDC, Soares AR, Ferrarese-Filho O. The effects of dopamine on root growth and enzyme activity in soybean seedlings. Plant Signaling & Behavior. 2013;**8**:1-7. DOI: 10.4161/psb.25477

[42] Soares AR, Marchiosi R, Siqueira-Soares RDC, Barbosa de Lima R, Dantas dos Santos W, Ferrarese-Filho O. The role of L-DOPA

in plants. Plant Signaling & Behavior. 2014;**9**:1-7. DOI: 10.4161/psb.28275

[43] Mastan SA, Ramayya PJ, Naidu LM, Mallikarjuna K. Antimicrobial activity of various extracts of *Mucuna pruriens* leaves. Biomedical and Pharmacology Journal. 2009;**2**:55-60

[44] Champatisingh D, Sahu P, Pal A, Nanda G. Anticataleptic and antiepileptic activity of ethanolic extract of leaves of *Mucuna pruriens*: A study on role of dopaminergic system in epilepsy in albino rats. Indian Journal of Pharmacology. 2011;**43**:197-199. DOI: 10.4103/0253-7613.77368

[45] Sree V, Aswani V, Babu R, Kumar P, Varma V, Birudu R. Advancements in the production of secondary metabolites. Journal of Natural Products Review. 2010;**3**:112-123

[46] Oksman-Caldentey KM, Inzé D. Plant cell factories in the post-genomic era: New ways to produce designer secondary metabolites. Trends in Plant Science. 2004;**9**:433-440. DOI: 10.1016/j.tplants.2004.07.006

[47] Smetanska I. Production of secondary metabolites using plant cell cultures. Advances in Biochemical Engineering/ Biotechnology. 2008;**111**:187-228. DOI: 10.1007/978-3-540-70536-9

[48] Hussain MS, Rahman MA, Fareed S, Ansari S, Ahmad I, Saeed M. Current approaches toward production of secondary plant metabolites. Journal of Pharmacy & Bioallied Sciences. 2012;**4**:10. DOI: 10.4103/0975-7406.92725

[49] Lalasangi SI, Hiremath L. Regeneration elicitation of bioactive compound extracted from endangered medicinal plant Mucuna pruriens and study of its inhibition activity using biomolecular simulation studies. Proj. Ref. No. 40S _ BE _ 2138. n.d

[50] Chattopadhyay S, Datta SK, Mahato SB. Production of L-DOPA from cell suspension culture of *Mucuna pruriens* f. *pruriens*. Plant Cell Reports. 1994;**13**:519-522

[51] Janarthanam B, Sumathi E. Optimization of biomass culture yield and l-dopa compound in the callus culture from cotyledonary leaves of *Mucuna pruriens*. Asian Journal of Pharmaceutical and Clinical Research. 2015;**8**:4-8

[52] Adam JI, Dampare SB, Addae G. Effect of culture conditions on L-dopa accumulation in callus culture of *Mucuna pruriens*. Journal of Chemical and Pharmaceutical Research. 2010;**2**:504-527

[53] Namdeo AG. Plant cell elicitation for production of secondary metabolites: A review. Pharmacognosy Reviews. 2007;**1**:69-79. DOI: 10.1016/ S0168-9452(01)00490-3

[54] Patel H, Krishnamurthy R. Elicitors in plant tissue culture. Journal of Pharmacognosy and Phytochemistry. 2013;**2**:60-65

[55] Simic SG, Tusevski O, Maury S, Delaunay A, Joseph C, Hagège D. Effects of polysaccharide elicitors on secondary metabolite production and antioxidant response in *Hypericum perforatum* L. shoot cultures. Scientific World Journal. 2014;**2014**:1-10. DOI: 10.1155/2014/609649

[56] Gadzovska S, Maury S, Delaunay A, Spasenoski M, Hagège D. The influence of salicylic acid elicitation of shoots, callus, and cell suspension cultures on production of naphtodianthrones and phenylpropanoids. Plant Cell Tissue and Organ Culture. 2014;**113**:25-39. DOI: 10.1007/s11240-012-0248-0

[57] Wong SP, Leong LP, William Koh JH. Antioxidant activities of aqueous extracts of selected plants. Food Chemistry. 2006;**99**:775-783. DOI: 10.1016/j.foodchem.2005.07.058

[58] Raghavendra S, Kumar V, Ramesh CK, Khan MHM. Enhanced production of L-DOPA in cell cultures of *Mucuna pruriens* L. and *Mucuna prurita* H. Natural Product Research. 2012;**26**:792-801. DOI: 10.1080/14786419.2011.553721

[59] Longhi JG, Perez E, de Lima JJ, Cândido LMB. In vitro evaluation of *Mucuna pruriens* (L.) DC. antioxidant activity. Brazilian Journal of Pharmaceutical Science. 2011;**47**:535-544. DOI: 10.1590/S1984-82502011000300011

[60] Murthy SN, Malgaonkar MM, Shirolkar AR, Pawar SD, Sangvikar S, Kulkarni YR. A comparative assessment of pharmacologically active principles and antioxidant activity of commonly occurring Mucuna SPS. in India. International Journal of Ayurveda and Pharma Research. 2015;**3**:8-13

[61] Yadav SK, Rai SN, Singh SP. *Mucuna pruriens* reduces inducible nitric oxide synthase expression in Parkinsonian mice model. Journal of Chemical Neuroanatomy. 2017;**80**:1-10. DOI: 10.1016/j.jchemneu.2016.11.009

[62] Oseni OM, Pande V, Nailwal TK. A review on plant tissue culture, a technique for propagation and conservation of endangered plant species. International Journal of Current Microbiology and Applied Sciences. 2018;**7**:3778-3786. DOI: 10.20546/ijcmas.2018.707.438

[63] Mohan VR. Antibacterial activity of *Mucuna pruriens* (L.) Dc. var. pruriens—An ethnomedicinal plant. Science Reserch Reporter. 2011;**1**:69-72. Available from: http://jsrr.in

[64] Rayavarapu KA, Kaladhar DSVGK. Evaluation of antimicrobial activity of mucuna pruriens on plant pathogens. Asian Journal of Biochemical and Pharmaceutical Research. 2014;**1**:593-600

[65] Pandey J, Pandey R. Study of phytochemical and antimicrobial activity of alcohlic extract of *Mucuna pruriens* (L.) leaves. International Journal of Applied Research. 2016;**2**:219-222

[66] Houghton PJ, Osibogun IM. Flowering plants used against snakebite. Journal of Ethnopharmacology. 1993;**39**:1-29. DOI: 10.1016/0378-8741(93)90047-9

[67] Kumar A, Gupta C, Nair DT, Salunke DM. MP-4 contributes to snake venom neutralization by *Mucuna pruriens* seeds through an indirect antibody-mediated mechanism. The Journal of Biological Chemistry. 2016;**291**:11373-11384. DOI: 10.1074/jbc.M115.699173

[68] Gómez-Betancur I, Gogineni V, Salazar-Ospina A, León F. Perspective on the therapeutics of anti-snake venom. Molecules. 2019;**24**:1-29. DOI: 10.3390/molecules24183276

[69] Guerranti R, Aguiyi JC, Neri S, Leoncini R, Pagani R, Marinello E. Proteins from *Mucuna pruriens* and enzymes from Echis carinatus venom. Characterization and cross-reactions. Journal of Biological Chemistry. 2002;**277**:17072-17078. DOI: 10.1074/jbc.M201387200

[70] Kasturiratne A, Wickremasinghe AR, De Silva N, Gunawardena NK, Pathmeswaran A, Premaratna R, et al. The global burden of snakebite: A literature analysis and modelling based on regional estimates of envenoming and deaths. PLoS Medicine. 2008;**5**:1591-1604. DOI: 10.1371/journal.pmed.0050218

[71] Shekins O. Anti-venom activity of *Mucuna pruriens* leaves extract against cobra snake (*Naja hannah*) venom. International Journal of Biochemistry Research & Review. 2014;**4**:470-480. DOI: 10.9734/ijbcrr/2014/10394

[72] Fung SY, Tan NH, Liew SH, Sim SM, Aguiyi JC. The protective effects of *Mucuna pruriens* seed extract against histopathological changes induced by Malayan cobra (*Naja sputatrix*) venom in rats. Tropical Biomedicine. 2009;**26**:80-84

[73] DeMaagd G, Philip A. Parkinson's disease and its management. Part 1: Disease entity, risk factors, pathophysiology, clinical presentation, and diagnosis. Pharmacology & Therapeutics. 1994;**40**:504-510. DOI: 10.1136/bmj.308.6923.281

[74] Katzenshlager R, Evans A, Manson A, Palsalos PN, Ratnaraj N, Watt H, et al. *Mucuna pruriens* in Parkinson's disease: A double blind clinical and pharmacological study. Journal of neurology, neurosurgery, and psychiatry. 2004;**75**:1672-1677. DOI: 10.1136/jnnp.2003.028761

[75] Lieu CA, Venkiteswaran K, Gilmour TP, Rao AN, Petticoffer AC, Gilbert EV, et al. The antiparkinsonian and antidyskinetic mechanisms of *Mucuna pruriens* in the MPTP-treated nonhuman primate. Evidence-Based Complementary and Alternative Medicine. 2012;**2012**. DOI: 10.1155/2012/840247

[76] Kasote DM, Katyare SS, Hegde MV, Bae H. Significance of antioxidant potential of plants and its relevance to therapeutic applications. International Journal of Biological Sciences. 2015;**11**:982-991. DOI: 10.7150/ijbs.12096

[77] Waris G, Ahsan H. Reactive oxygen species: Role in the development of cancer and various chronic conditions. Journal of Carcinogenesis. 2006;**5**:1-8. DOI: 10.1186/1477-3163-5-14

[78] Koller WC, Rueda MG. Mechanism of action of dopaminergic agents in Parkinson's disease. Neurology. 1998;**50**. DOI: 10.1212/wnl.50.6_suppl_6.s11

Chapter 9

Hedysarum Species from Caucasus

Serebryanaya Fatima and Imachueva Djavgarat

Abstract

In the complex pharmacognostic studies of three species of the genus *Hedysarum* L., we have developed a method for quantitative determination by UV spectrophotometry, HPLC, and capillary electrophoresis of the sum of xanthones in terms of mangiferin. The technique takes into account the basic physicochemical properties of xanthones; is characterized by reproducibility, high accuracy, and simplicity; and allows conducting both a screening assessment of various raw materials containing mangiferin derivatives and standardization of the prepared vegetable raw materials. The developed methods are tested on the aboveground organs of *Hedysarum* species, which are collected and dried by taking into account the rules and requirements for the preparation of medicinal raw materials. As a result, it was found that the greatest quantitative content of the sum of xanthones in terms of mangiferin is distinguished by the grass *H. caucasicum* M. Bieb. (0.62 ± 0.021%). For the first time, morphological-anatomical diagnostic signs of the species *Hedysarum caucasicum M. Bieb.*, *Hedysarum daghestanicum Rupr. ex Boiss.*, *Hedysarum grandiflorum Pall.* are necessary for standardization of medicinal vegetal raw materials. The results show the prospect of further investigation of *Hedysarum* as an additional source of mangiferin.

Keywords: *Hedysarum*, xanthones, mangiferine, antiviral therapy

1. Introduction

To improve modern pharmacy, the study of biologically active substances exhibiting specific pharmacological activity from the natural raw materials of plant origin is of obvious interest. In our time, the problem of obtaining antiviral and antibacterial agents of plant origin is especially acute. At the moment, the problem of chronic recurrent viral diseases is very relevant. According to WHO data, about 67% of the world's population suffer from diseases of various organs and systems caused by the Herpes simplex virus.

Mangiferin (2-C-β-D-glucopyranosyl-1,3,6,7-tetraoxyxanthone) belongs to the group of xanthones proper, is the most widespread representative of C-glycosides, and has antiviral and antibacterial properties (**Figure 1**).

In species of the genus *Hedysarum*, Fabaceae, xanthones, namely mangiferin, are the main group of biologically active substances. Russian scientists have developed an antiviral drug Alpizarin, from *H. alpinum* L. and *H. flavescens* Regel & Schmalh, the active substance of which is mangiferin, which has an antiviral activity against DNA-containing viruses (*Herpes simplex, Varicella zoster, Cytomegalovirus*), immunostimulating properties, and a bacteriostatic effect on Gram-positive and Gram-negative bacteria. VILAR (Russia) produces this in two forms of production "Alpizarin ointment" and "Alpizarin tablets." In addition to domestic drugs, the

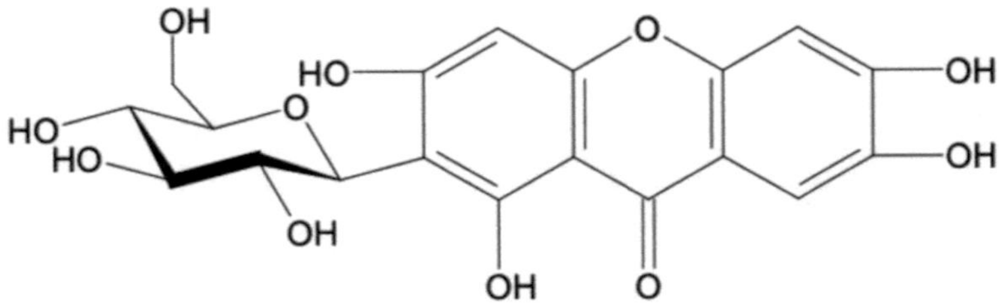

Figure 1.
Mangiferin (2-C-β-D-glucopyranosyl-1,3,6,7-tetraoxyxanthone).

foreign antiviral drug Mangogerpin (BV Pharma, Vietnam) is known, which also has two forms of release: tablets and ointment, the active substance of which is mangiferin.

There is a need to expand the raw material base for the production of broad-spectrum antiviral drugs. In this regard, a comprehensive analysis, taking into account the features of growth, biological characteristics, chemical composition, and known pharmacological properties, will make it possible to substantiate the directions of their rational use as well as the creation of additional domestic raw materials of mangiferin as a known antiviral drug based on medicinal plant raw materials. Thus, the study of species of the following genus is promising: *Hedysarum* L., *Hedysarum caucasicum* M. Bieb., *Hedysarum grandiflorum* Pall., and *Hedysarum daghestanicum* Rupr. ex Boiss., which are growing in the North Caucasus as additional raw materials for mangiferin.

The purpose of our work is to study plants of species of the genus *Hedysarum* L., including *Hedysarum caucasicum* M. Bieb., *Hedysarum daghestanicum* Rupr. ex Boiss., and *Hedysarum grandiflorum* Pall., as additional raw materials for mangiferin.

To achieve this goal, it is advisable to achieve the following objectives:

- Conduct a comparative morphological-anatomical study of these species of the genus *Hedysarum L.* (*Hedysarum caucasicum M. Bieb.*, *Hedysarum daghestanicum Rupr. ex Boiss.*, and *Hedysarum grandiflorum* Pall.) and identify the main morphological-anatomical diagnostic features.

- Develop an effective procedure for quantifying the sum of xanthones in terms of mangiferin from these species of the genus *Hedysarum* L., by methods of UV spectrophotometry, capillary electrophoresis, and high-performance liquid chromatography.

The family *Fabaceae Lindl.* (legumes) has about 650 genera and over 18,000 species, whose range of growth covers all continents of the globe. Within its range, species of the legume family are characterized by a variety of habitats. Life-form of members of the family *Fabaceae* Lindl.—perennial herbaceous plants, semiplants, and rarely trees—there are tree or herbaceous lians [1]. The leaves are complex with sills. The leaflet is next. Flowers are in finite or sinus brushes. The flowers are zygomorphic, quite large, and bright. The near-wind is double, the cup is most often actinomorphic, consists of five converged cups, and the crown consists of five petals, two of which grow into a "boat," one large—a "sail" and two small petals —"vests" remain free. The android consists of 10 stamens, nine of which usually grow together and one stamen remains free. Gynecium monocarpic tie the upper one. The fruit of legumes is monocarpium, one-, two-, or multifamily legumes, very diverse in morphological and anatomical features. Seeds are without endosperm or with scant endosperm. Spare nutrients are deposited directly in the seed.

The genus *Hedysarum* L. combines about 285 species, which are predominantly common in Eurasia. The genus *Hedysarum* is distributed mainly in the European part, but some species are also found in Asia. *H. caucasicum* Bieb. grows in all highland areas of the North Caucasus, including the Rocky Ridge, in the alpine belt, up to 3500 m [2]. This species is endemic, growing on the highland meadows of the Caucasus: in the pre-Caucasus, Western and Eastern Transcaucasia, in Daghestan. It is described from the Caucasus type in Leningrad [1]. C. Linnaeus (1753) and B. A. Fedchenko (1902) were engaged in cystematics of the species of *Hedysarum* L. The species *Hedysarum caucasicum* Bieb we are investigating first was described in 1808 [3]. In the flora of Caucasus in 1873, the species was classified as *H. obscurum var. caucasicum* Trautv. The origin of the name of this plant is related to the shape of the fruit. Beans in appearance resemble coin meetings, and the Latin name *Hedysarum* comes from the Greek words "hedys" aroma, which literally means pleasantly smelling. Members of this genus are perennial herbs, shrubs, or semiplants of seamless or stem-shortened plants. They grow both in forests and on raw meadows, on riverbanks, and on alpine and subalpine meadows, on rocky sprouts, and in steppes.

Stems are often highly branched, branching, and sometimes completely undeveloped and the flower arrow comes out of shortened shoots developing at the neck of the rhizome; leaves are unparalleled, common: 5–9-paired, less common: 1–3-paired or even consisting of just one unparalleled leaf (**Figure 2**). The genus is characterized by simple botrioid flowers—brushes; the cup is bell-shaped, its teeth are search-and-take longer than the tube; the crown exceeds the cup; and wings are slightly or two times shorter than a boat, less often longer than it. The beans are artenic, and sometimes part of the seed does not develop and the bob consists of 1–3 artens; sprouts are flat-compressed or slightly convex, smooth, bare, or more often dried, mesh or with transverse ribs, often shrunk with short or longer bristles.

According to Flora of the USSR, in the *Hedysarum caucasicum* Bieb.—a plant 30–60 cm high, the stems are straight or ascending, not shortened, olfactory; and the leaves are 7- are 12-pairs, elliptical or egg-like long-lasting-watt, with sharpness at the top, 12–16 mm long, 7–9 mm wide. Flowers (without brush) are longer than leaves; brushes are not very thick, of 25–35 flowers; the lower tooth of the cup is equal to the tube, the rest are shorter; and the crown is dark-magenta, 16–18 mm

Hedysarum caucasicum M.Bieb — *Hedysarum daghestanicum* Rupr. ex Boiss — *Hedysarum grandiflorum* Pall

Figure 2.
Species of Hedysarum *from Caucasus (Russia).*

long. The fruit is a bob of 3–6 arthropods, naked or desiccated; the mushroom of the bean is oblong-elliptical, and its edge is not wide.

The microstructure of the above-ground parts of some members of the *Hedysarum* was studied by Ladygina et al. [4]. The features of the anatomical structure of leaves of five species of *Hedysarum* from the section *Gamotion* Basin were studied. These species containing mangiferin have been found to have bright golden yellow fluorescence of epidermis cells, which can be used to detect mangiferin directly in plant material [5].

We did not meet the information on the study of anatomical features of the structure of the organs in the accessible literature. The onthogenesis of species *Hedysarum* usually includes four periods and 10 age states. In the example of *H. austrosibiricum*, age spectra can be divided into four types, which probably reflect ecological-phytocenotic growth conditions [6].

During 3 years of life, alpine individuals undergo the following age conditions: in the first year—seedlings, juvenile, immature, and more than 60% of individuals are adult vegetative; in the second year, more than 90% of individuals enter a young reproductive state; and in the third year, all individuals transition to a reproductive state. Seeds of tested plants need pre-sowing treatment. Mechanical scarification and treatment with concentrated sulfuric acid are the best ways to disrupt the rest of both species [6–8].

The flowering phase of the *Hedysarum caucasicum*—the beginning and the middle of July, and the fruiting phase—the end of July—and the beginning of August [1]. For the genus *Hedysarum*, the following set of chromosomes is characteristic: 14, 16, and 48. For *Hedysarum caucasicum*, the chromosome number is 14 [9].

2. Material and methods

The object of the study was the grass of *Hedysarum caucasicum* Bieb. family legumes (Fabaceae), which were collected in the flowering phase on the southeast slope of Mount Alibek at an altitude of 2200 m (Dombay District, Russia).

Freshly harvested and dried samples of raw materials of the species of the genus *Hedysarum* L., including *H. caucasicum* M. Bieb., *H. daghestanicum* Rupr. ex Boiss., and *H. grandiflorum* Pall. were obtained (**Table 1**). The collected test raw materials were dried without direct sunlight in the air.

Research methods: comparative morphological, ecological-geographical, molecular-genetic, morphological-anatomical and histochemical, and phytochemical (UV-spectrophotometry, HPLC, chromato-mass spectrometry, capillary electrophoresis).

Species	№	Geographical coordinates	Altitude above sea level	Ecological and biological features of growth
Hedysarum caucasicum M. Bieb.	1.	N—43°43′46″, E—42°53′65”	2100 m	Open moistened areas among large boulders
	2.	N—40°30′17″, E—44°40′0”	2200 m	Southeastern slope of Alibek, gorge of Mount Alibek
	3.	N—42°19′32,21″, E—47°09′49,1”	1250 m	Deep location of river valleys between mountain spurs
Hedysarum grandiflorum Pall.	4.	N—49°45′56″, E—44°16′16”	200 m	Deep location of river valleys between mountain spurs

Species	№	Geographical coordinates	Altitude above sea level	Ecological and biological features of growth
	5.	N—42°19′32,21″, E—47°09′49,1”	1250 m	Deep location of river valleys between mountain spurs
Hedysarum daghestanicum Rupr. ex. Boiss.	6.	N—42°38′15,7″, E—46°09′45,8”	850 m	Rocky-fine-brimmed steep slope of northeastern exposition
	7.	N—42°41′38,2″, E—46°14′27,2”	1100 m	The slope of the southeast exposition, steepness 30°
	8.	N—42°59′32,9″, E—46°54′46,9”	460 m	Steepened rocky-fine-crushed steep section at the top of the slope
	9.	N—42°19′32,21″, E—47°09′49,1”	1250 m	Deep location of river valleys between mountain spurs

Table 1.
List of objects of study with indication of ecological and biological features of growth.

3. Methods for quantitative analysis of mangiferin

High-performance liquid chromatography, photocolorimetry, complexonometry, and chromato spectrophotometry are most commonly used to quantify mangiferin content in plant raw materials as well as biological fluids [10–15]. The photocolorimetric definition of mangiferin in both the raw material and crystalline powder has a number of advantages, such as shortening the analysis steps and replacing the deficient vegetable, tetrahydrofuran, with dioxane. The method is based on mangiferin's ability to produce a complex compound with chlorine iron. Mangiferin was determined from a calibration plot of the solution optical density versus mangiferin concentration. When comparing spectrophotometric and photocolorimetric methods, almost the same results were obtained [5, 10, 12, 16, 17].

As for the current method of high-performance liquid chromatography, according to literature, separation of xanthone glycosides by HPLC method could not be achieved on sorbents containing amino and cyano groups [7, 18–23]. The best results were obtained on reverse phase C18 sorbents, and methanol-water, ethanol-water, and acetonitrile-acetic acid were used as the mobile phase. In aqueous systems, the shape of the mangiferin peak deteriorated. To quantify mangiferin in biological fluids, a sensitive reverse phase HPLC technique is proposed [17, 24, 25].

The mobile phase was acetonitrile and a 3% CH_3COOH solution at a ratio of 16:84 was chromatographed at a wavelength of 254 nm using an external standard method [11, 26].

Furthermore, in the quantitative determination of mangiferin by HPLC, a system consisting of acetonitrile, water, and phosphoric acid was used as the mobile phase. The selected conditions allowed to achieve a clear separation of mangiferin and isomangiferin peaks on chromatogram. In addition to the previous systems, methanol, tetrahydrofuran, or acetonitrile, an aqueous solution of phosphoric acid in various ratios are used as the mobile phase in HPLC gradient elution. Of the latest techniques for the quantitative determination of mangiferin, liquid chromatography followed by mass spectrometric determination was used, and the method is characterized by speed and quality [27–30].

The chromate-spectrophotometric examination is based on sequential chromatography and spectrophotometry. The raw material is treated with the following extractant system—acetone:water in ratio 1:1 with acidification with 5% hydrochloric acid followed by chromatography in the system with 15% acetic acid. The eluation of the zones with mangiferin is preparing after viewing the chromatogram in UV light. The optical density of the solutions was measured at 372 nm. The mangiferin content was calculated from the specific absorption index.

There are a number of intense absorption bands in the UV spectrum of mangiferin. The most convenient to quantify a substance is a band with a maximum at 369 nm and a specific absorption coefficient of 295 ± 0.92. In the field of working concentrations, the absorption of mangiferin solutions is subject to Lambert-Beer law [23].

Analysis of literary sources has shown that the genus *Hedysarum* L. combines about 285 species, which are predominantly common in Eurasia. *H. caucasicum* Bieb. is endemic, growing in the highland meadows of the Caucasus. In a literary search, it has been found that chemical study information relates mainly to *H. alpinum* and *H. flavescens*. Information on the chemical study of *H. caucasicum* is fragmented and insufficient. Plants of the genus *Hedysarum* are widely used in folk and waitinal medicine, as an antibacterial, antiviral, immunomodulatory, and anti-inflammatory agent [6, 8, 17, 26, 31–39]. The following methods of analysis are used to identify and quantify the main active substances of xanthones: spectrophotometry, photocolorimetry, and chromatographic methods of analysis (TLC, HPLC, and mass spectrometry) [40–42]. The study of theoretical bases of extraction of medicinal raw materials allows to find the optimal conditions of technology for creation of medicinal forms on the basis of herbal of Caucasus.

4. Results

Morphological study: Life-form—a perennial herbaceous plant 40–50 cm high, and underground organs reach 30 cm length. The escape is elongated, branched, straight-standing, or raised. The number of leaflets located on rachis varies from 11 to 15. The leaves have an egg shape, a rounded base, a whole edge, and a spiky top. Decaying of leaflets is insignificant. The flower is simple botriode, brush. The flower is zygomorphic. The cup consists of five cups, a moth-type crown, and the color of the petals is pink-purple. The android consists of nine converged tangles and one free. Ginecey is monocarpal. Tie the top one. Fruit: according to morphogenetic classification of fruits, it refers to monocarp; according to morphological—bob, flat, oblong in shape, and consists of rounded arthropods. Number of squads range from 3 to 5.

The leaf is amphistomatic. The upper epidermal has weak and almost straight anticline cell walls. The mouth is abnormal, surrounded usually by 3–5 near-oral cells. Trichomes were not detected on the abaxial side of the leaf. The lower epidermal has strong anticline cell walls. An anomocytic type oyster is surrounded by usually 3–5 parotid cells. The embossing is formed by simple single-cell hairs located either in the region of the veins or on the edge of the leaf plate. When considering the sheet microreparation from the surface, rhombic calcium oxalate crystals located in large veins are found. They form a characteristic crystalline lining of the veins. On the cross section, the leaves have a characteristic dorsoventrale structure (**Figure 3**).

The palisade mesophyll localizes only under the upper epidermis in one layer, its cells are stretched, tightly pressed to each other. The mechanical fabric is a collenchym which is located both under the upper and lower epidermis in the region of the core. In the central part of the main core, there is a large collateral conducting bundle, and from the dorsal side to the conducting bundle there is a

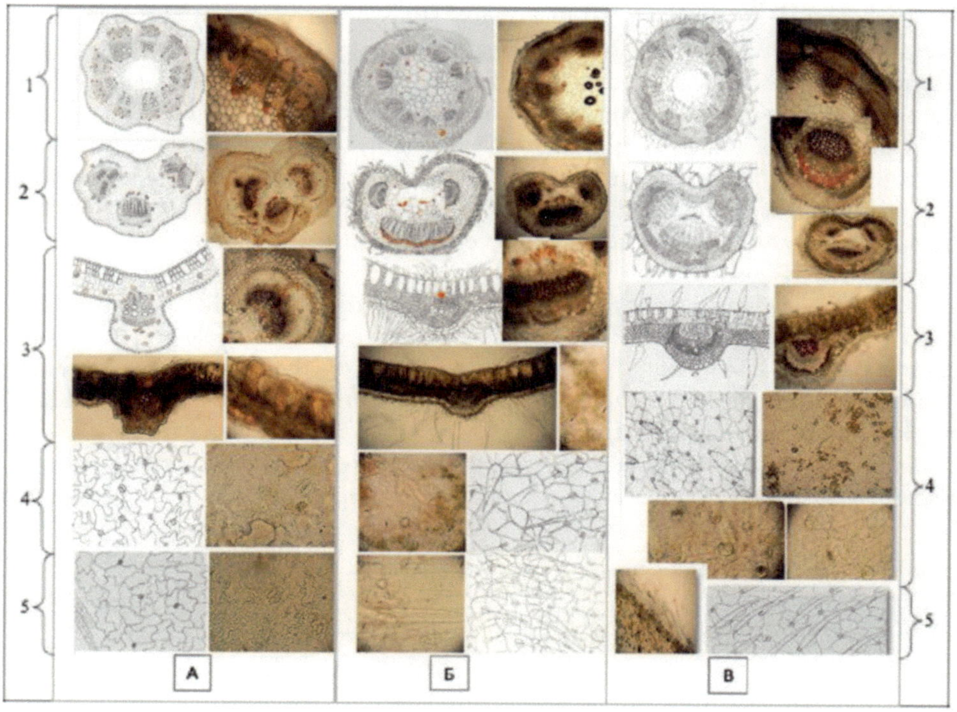

Figure 3.
Microscopic signs of the above-ground organs of Hedysarum caucasicum *M. Bieb. (A)* Hedysarum daghestanicum *Rupr. ex Boiss. (B)* Hedysarum grandiflorum *Pall. (C) 1. Transversal section of the stem in the lower part; 2. transversal section of the leaf petiole in the lower part; 3. transverse slice of leaf plate; 4. the adaxial epidermis of the leaf blade; and 5. the abaxial epidermis of the leaf blade.*

sclerenchym of pericyclic origin. The leaflets are arranged on small cherries, which have cylindrical shape on the cross section. The covering tissue of the cherry is represented by an epidermal, under it there is a collenchym in 1–2 layers, and the main volume occupies chlorenchyma. The collateral conducting bundle surrounded by sclerenchym is located in central part of cut. The rachis at the bottom has a grooved shape with a deep horseshoe spoon on the abaxial side. Collateral conducting bundles of 6–8 are arranged in an arc-like manner. It is interesting that idioblasts with yellow content are found in the phloem, in the pericyclic zone, and in the parenchyma of the core adjacent to the xylem portion of the conductive beam.

The stem has a polyhedral shape on the cross section. A cavity may form in the central part of the stem. Under the epidermal continuous ring in 1–2 layers of cells, there is a plate-type collenchym. In the projection regions, the number of layers of collenchym cells increases to 5–7. Chlorenchym is located following collenchym in discontinuous regions in 2–3 layers of cells. The pericyclic zone is represented by sclerenchymal fibers arranged in discontinuous areas above the conducting beams. Conducting system is of bundle type. Conducting beams are of open collateral type in an amount of 14–16 (**Figure 3**).

The cambium forms secondary conductive beams and thick-walled parenchyma, which is subsequently strongly ligated. Idioblasts are found, as in the cherry, in the phloem part of the conducting beam, in the pericyclic zone, as well as in the parenchyma of the core. The abovementioned micromorphological characteristics of the above-ground organs of the *Hedysarum caucasicum* can be used in drawing up regulatory documentation for the Hedysarum caucasicum grass as an additional raw material source of xanthone glycoside mangiferin.

A comparative morphological-anatomical study of three species of the genus *Hedysarum* L., growing in the North Caucasus, was carried out. The most significant diagnostic morphological features have been established, which provide to establish the species belonging of the medicinal raw materials specified in **Table 2**.

A comparative analysis of the anatomical structure of the stem revealed common features, which include the multifaceted shape of the stem on the cross section, and the pubescence is represented by simple bicellular hairs, which have a narrowed base and an expanded structure, a characteristic feature of the hairs is the wariness of the cuticle (**Figure 3**).

In a comparative study of the anatomical structure of the caulifoliar system, the main attention was drawn to the anatomical structure of the petiole and rachis of a complex leaf. The following differences were revealed—the shape on the cross section of the petiole of a complex sheet changes from a clearly saddle-shaped to a rounded-triangular shape, the number of ribs on the abaxial side changes.

Differences were identified regarding the structure and location of collenchyma, sclerenchyma, and the conducting system. The conductive system of the petiole and

Species	Life-form	Leaves	Flowers	Fruits
Hedysarum caucasicum M. Bieb.	High plant, 30–50 cm tall. The lower internodes are not shortened	The leaves are scattered throughout the stem, from 7 to 12 pairs of elliptical or ovate oblong leaflets with a pointed tip at the apex. Pubescence is not developed	The brushes on long legs, 11/2–2 times longer than the leaves, are not very thick. The lower tooth of the calyx is equal to the tube, the rest are shorter. Dark-purple petals	The segments are not large, smooth, or slightly toothed
Hedysarum grandiflorum Pall.	Perennial 20–40 cm tall. Stemless or almost stemless. Stipules are large, membranous, fused, brown, and scattered-hairy	The leaves are petiole, short pressed-hairy and, long spaced pubescent. Leaflets 1–4-paired, ovate, or broadly elliptical, and large, slightly hairy above, less often glabrous, densely silvery-silky below	The brushes are multi-flowered, with deflected flowers. Bracts are lanceolate, light brown, hairy, bell-shaped calyx, its teeth are linear-awl-shaped, densely thawed-hairy, Corolla yellow	Beans 2–5-segmented, segments rounded, densely white-haired, mesh-ribbed, along the edges with spines bent inside
Hedysarum daghestanicum Rupr. ex Boiss.	Stem-free rod-root perennial. Peduncles with brushes 10–25 cm long. All parts of the plant have a grayish plaque	Ten leaves with 3–4 pairs of leaflets, leaflets oblong, up to 18 mm long and up to 8 mm wide. The apical leaf is larger. Accrete stipules	The brushes are small-flowered, thick. The flowers are large, creamy white or purple. The calyx is four times shorter than the corolla. The boat is shorter than the flag and two times longer than the wings. The flowers are large, cream-white	Beans from 2 to 4 segments. Lentil-shaped, warty segments

Table 2.
Comparative morphological characterization of Hedysarum *L. species growing in the North Caucasus.*

rachis of the complex sheet is represented by three large and 2–7 small conducting bundles (**Figure 3**).

All conducting bundles have a pronounced collateral structure, while one large, dorsal conducting bundle is located on the adaxial side, and it has a characteristic rounded or semilunar shape and is reinforced with sclerenchymal fibers on the phloem side. Two ventral conducting bundles are smaller and are usually located in the rib zone, also reinforced by sclerenchyma on the phloem side.

Thus, the rachis zone is usually characterized by 2–4 additional conductive beams, or they are completely absent. The number and arrangement of conducting beams seem to us most interesting and more informative from the point of view of diagnostics of this species, growing under various environmental conditions (**Figure 1**). When conducting a comprehensive morphological-anatomical study of three species of the genus *Hedysarum* L., the morphological-anatomical diagnostic features were revealed, which are presented in this chapter.

5. Determination of humidity

By the moisture content of the raw material, it is meant a loss in mass due to hygroscopic moisture and volatile substances, which is determined by drying to a constant mass. Medicinal plant raw materials should not contain moisture above permissible standards. Moisture content of the analyzed raw material was determined by the method of drying by Pharmacopea XIV [14]. The weight loss on drying was 8.60% (average of two parallel determinations).

6. Determination of total ash

The ash of plant raw materials refers to the residue of inorganic substances obtained after burning the raw materials and then calcining the residue to a constant weight. Plant ash (total ash) consists of a mixture of various inorganic substances in the plant itself and mineral impurities (earth, sand, dust, and stones) that can enter the raw materials when collected and dried. The common ash most commonly contains the following elements: Na, K, Ca, Mg, Fe, Si, F, P, and C, which are in the form of oxides or salts of carbonic, phosphoric, sulfuric, and other acids. Ash determination of total (x) was carried out by Pharmacopea of Russia. The total ash content was 4.04% (average of two parallel determinations).

7. Microbiological purity test

For research, the grass of the *Hedysarum caucasicum* Bieb. family legumes (*Fabaceae*) were collected in the flowering phase on the southeast slope of Mount Alibek (Dombay Gorge District, CHR) and used. According to amendment No. 3 to the article of Pharmacopea of the publication "Methods of microbiological control of medicines," introduced on June 19, 2003, the studied plant raw materials belong to the category 4Б—medicinal plant preparations and medicinal plant raw materials "angro," prepared without the use of boiling water. The requirements for this category are as follows:

- Total number of aerobic bacteria—no more than 105 in 1 g or 1 ml.
- Total number of fungi—not more than 104 in 1 g or in 1 ml.

- Enterobacteria and other gram-negative bacteria—not more 103 1 g or 1 ml.
- Absence of *Escherichia coli*—in 1 g or 1 ml.
- Absence of *Salmonella* in 10 g or 10 ml.

Before preparing the dosage form, the studied vegetable raw material of the *Hedysarum caucasicum* Bieb. was tested for microbiological purity. The presented results make it possible to conclude that according to the indicator "microbiological purity," the sample of vegetable raw materials of the *Hedysarum caucasicum Bieb.*, presented for analysis, meets the requirements for medicinal vegetable raw materials "angro," used without thermal treatment.

8. High-quality phytochemical analysis

Determination of tanning substances: About 1.0 g of raw material was poured with 100.0 g of water, heated for 20–30 min in a water bath and filtered. The following reactions were carried out with the resulting solution [15]: Several drops of iron ammonium alum were added to 2 ml of the solution, and black and green staining appeared, indicating the presence of condensed tanning agents; a few drops of a 1% solution of quinine hydrochloric acid were added to 2 ml of the solution, and opalescence appeared.

Determination of polysaccharides: For qualitative detection of polysaccharides, water extraction was prepared from 2.0 g of alpine penny roots in a water bath for 30 min. Then it was filtered off and the filter was washed with hot water. The recovery was evaporated to 1/5 volume and three times the volume of 96% ethanol was added thereto. As a result, a loose curd precipitate of the polysaccharide complex was formed. The precipitate was separated, dissolved in water, reprecipitated, washed with alcohol, and dried. The obtained polysaccharide complex is an amorphous mass soluble in water. In the composition of water-soluble polysaccharides of the penny, mucous substances predominate [43].

Definition of the restoring sugars: About 1.0 of the milled raw material was placed in a 25 ml flask, poured with 10 ml of water, and refluxed for 0.5 h. The solution was filtered through gauze and washed with water. 5 ml of the resulting solution was transferred to a tube and 15 ml of 95% ethyl alcohol was added. Precipitation of the bulk precipitate was observed. The solution was filtered, the precipitate was transferred to a test tube, 5 ml of diluted hydrochloric acid was added, boiled for several minutes, 5 ml of Feling reagent was added and boiled again, and orange-red staining was observed [14].

Determination of free organic acids: A 1:10 decoction was prepared from Caucasus penny grass while heating in a water bath for 1 h. Broth was filtered. Five drops of digestion were placed in the tube and adjusted to 1 ml with purified water. One drop of the methyl red indicator was added, and red staining was observed, indicating the presence of organic acids [15].

Definition of amino acids: For qualitative detection of amino acids, reaction with 0.1% solution of ninhydrin in n-butanol on filter paper was used, and characteristic blue-violet staining appeared in formation of Rueman complex [15].

Determination of flavonoids and xanthones: In order to determine flavonoids, it was necessary to obtain an alcohol extract from the raw material. Extraction was carried out with 80% ethyl alcohol. About 1 g of the feed was placed in a 25 ml flask, 10 ml of 80% ethyl alcohol was added and heated in a water bath for 10–15 min under reflux. The resulting solution was filtered through a paper filter after cooling. Reactions were carried out with the resulting solution.

Cyanidine sample: 0.1 g of magnesium dust, 2 ml of concentrated hydrochloric acid were added to 1 ml of extraction, heated in a water bath for 2–3 min, and after some time red-orange staining was observed; 2 drops of 2% basic lead acetate solution were added to 1 ml of the recovery, and yellow-lemon staining appeared; 1 ml of a 10% ammonia solution was added to 1 ml of the recovery, and yellow staining turned orange on heating appeared. 1 ml of a 2% solution of aluminum chloride in 96% ethyl alcohol was added to 1 ml of the recovery, and lemon-yellow staining was observed [31].

Qualitative reactions with these reagents showed the presence of flavonoid substances in the grass of the Caucasus penny, which allowed us to use the chromatography method for further analysis, which is widely used for their detection and identification. Chromatographic separation of the sum of flavonoids and xanthones was carried out in the preparation of the extracts, and ethyl alcohol of 96, 80, 60, and 40% concentration was used as the extractant. 0.05 ml of the *Hedysarum caucasicum* Bieb. extracts were applied to a 40 × 40 cm Whatman chromatographic paper and subjected to ascending chromatography in a solvent system: butanol-glacial acetic acid-water in a ratio of 4:1:5 compared to witness substances. When viewing the chromatogram in UV light, three main spots were found in extracts of the following concentrations of ethanol 96:80:60%. First spot corresponds to mangiferin, second to hyperoside, and third to campferol. Further, chromatograms were sprayed with alcohol solution $AlCl_3$, and a change in stain color was observed.

9. Qualitative detection and quantification of xanthones

Thin layer chromatography was used for qualitative detection of mangiferin. Chromatography was carried out in systems: n-butanol-acetic acid-water (4:1:5); chloroform-methanol-water (13:7:2); and 15% acetic acid, on "Sorbfil PTCC-AF-A" plates 10 × 10 cm and 10 × 15 cm.

The development of the plates was carried out by spraying with the following reagents: a solution of iron(III) chloride of 2%, an alcoholic solution of aluminum chloride of 1%, and ammonia vapors and UV radiation (fluorescent lamp UV-A). As a result of the TLC study of the extraction of raw materials there are seventy percent ethyl alcohol compared to standard samples indicated that xanthone glycoside-mangiferin was present in these samples (**Table 3**).

10. Quantification of the sum of xanthones mangiferin by UV spectrophotometry

The content of the sum of xanthones in the test subjects in terms of mangiferin was calculated in two ways, using the optical density of the solution of the standard sample of mangiferin and the value of the specific absorption index of mangiferin under similar conditions. *Determination of specific value of mangiferin uptake:* A precise suspension of a standard mangiferin sample (about 0.01 g) was placed in a measuring flask with a capacity of 25 ml, 20 ml of 70% ethyl alcohol was added, stirred until the standard sample was completely dissolved, and the volume of solution was adjusted to a mark in the flask. Aliquots from the resulting solution were placed in measuring flasks with a capacity of 25 ml and labeled with the same solvent. The optical density was measured on a spectrophotometer at a wavelength

Value R_f	**Color of chromatographic area**									
	***H. caucasicum* M. Bieb.**		***H. grandiflorum* Pall.**		***H. daghestanicum* Rupr. ex Boiss.**		**Standard mangiferin (Sigma-Aldrich)**		**Alpizarin**	
	Before detecting	**After detecting**	**Before detecting**	**After detecting**	**Before detecting**	**After detecting**	**Before detecting**	**After detecting**	**Before detecting**	**After detecting**
System 1: n-butanol-acetic acid-water (4:1:5)										
0.51	Pale yellow	Bright yellow	Pale yellow	Bright yellow	Pale yellow	Bright yellow	Pale yellow	Bright yellow	Pale yellow	Bright yellow
System 2: chloroform-methanol-water (13:7:2)										
0.56	Pale yellow	Bright green	Pale yellow	Bright green	Pale yellow	Bright green	Pale yellow	Bright green	Pale yellow	Bright green
System 3: acetic acid 15%										
0.38	Pale yellow	Bright yellow	Pale yellow	Bright yellow	Pale yellow	Bright yellow	Pale yellow	Bright yellow	Pale yellow	Orange

Table 3.
Results of chromatography of Hedysarum *L. species together with the standard mangiferin (Sigma-Aldrich).*

of 365 nm in the resulting solutions. Ethyl alcohol 70% was used as a comparison solution. **Tables 4** and **5** shows the results of the experiment.

The obtained results were statistically processed; the relative error of determination was 1.4%, which makes it possible to conclude the reliability of the obtained results. The optical density of solution B was measured on a spectrophotometer at a wavelength of 365 ± 2 nm in a cuvette with a layer thickness of 10 mm. As the comparison solution, 70% ethyl alcohol was used. The quantitative content of the sum of xanthones in terms of mangiferin was calculated using the optical density value of the standard sample of mangiferin(I) and the value of the specific absorption index of mangiferin(II) established by us experimentally. **Table 5** shows the results of quantification of the sum of xanthones in terms of mangiferin in the raw material (**Table 5**).

As a result of complex chromatographic studies of three species of the genus *Hedysarum*, it was revealed that the maximum content of the sum of xanthones in terms of mangiferin is 0.62 ± 0.021%, and this is observed in the grass of *Hedysarum* (**Table 6**). The obtained results indicate the prospect of further study of the above-ground part, which allows us to consider this species as an additional raw material source of mangiferin.

Aliquot, ml	Concentration of mangiferin, % $\times 10^{-4}$	Optical density, *A*	Calculated value of specific key figure absorption, $A_{1CM}^{1\%}$	Metrological characteristics
0.25	4	0.1278	319.5	$S_{\bar{x}}$=18,433 $\Delta X = 4.52$ $t_{0,95} = 2.45$ 325.1 ± 4.52 $\varepsilon = 1.4\%$
0.50	8	0.2603	325.4	
0.75	12	0.3945	328.8	
1.00	16	0.5178	325.7	
1.25	20	0.6427	321.4	
1.50	24	0.7718	321.5	
1.75	28	0.9338	333.5	
			$\bar{A}_{1CM}^{1\%}$=325.1	

Table 4.
Results of determination of specific value of mangiferin absorption (exact weight 0.0101 g).

Mass, g	Value of optical density, A_x (λ = 364 nm)	Maintenance of the sum of xanthones, % (A_0)	Maintenance of the sum of xanthones, % ($A_{1cm}^{1\%}$)	Metrological characteristics	
		I	II	I	II
10.015	0.4621	0.640	0.637	$\acute{X}$ =0.629	$\acute{X}$ = 0.624
10.018	0.4672	0.647	0.644	$S_{\acute{x}}$ =0.0830	$S_{\acute{x}}$ =0.0830
10.100	0.4671	0.642	0.639	Δx =0.021	Δx =0.021
0.9989	0.4329	0.601	0.599	$\acute{x} \pm \Delta x$ =	$\acute{x} \pm \Delta x$ =
0.9996	0.4358	0.605	0.602	0.629 ± 0.021	0.624 ± 0.021
0.9898	0.4504	0.632	0.629	ε =3.39%	ε =3.42%

Table 5.
Content of the sum of xanthones in terms of mangiferin (a_o = 0.0101 г; A_o = 0.5178; $A_{1CM}^{1\%}$ =325.1; w = 7.23%).

Species	The content of the sum of xanthones in terms of mangiferin, %
Hedysarum caucasicum M. Bieb.	0.62 ± 0.02
Hedysarum grandiflorum Pall.	0.60 ± 0.02
Hedysarum daghestanicum Rupr. ex Boiss.	0.56 ± 0.01

Table 6.
The content of the sum of xanthones in terms of mangiferin in the above-ground part of species of the genus Hedysarum *L. by the value of the specific absorption index of mangiferin.*

11. Quantification of mangiferin in grass of *Hedysarum* species by capillary electrophoresis

For the study of these species of the genus *Hedysarum* capillary electrophoresis "Kapel - 105m" (Lumex Marketing OJSC, Russia), quartz capillary (L_{eff}/L_{tol} = 50/60 cm, ID = 75 μm) was used. The quartz capillary was previously washed successively with purified water, 1 M aqueous solutions of sodium hydroxide and hydrochloric acid.

The optical density of the prepared solution was measured on a spectrophotometer in the wavelength range of 200–500 nm. For this purpose, an aliquot of 1 ml was placed in two measuring flasks with a capacity of 25 ml. In one of the measuring flasks, the solutions were labeled with alcohol 70% ethyl, and in the other with borate buffer solution 0.01 M. As comparison solutions, ethyl alcohol 70% was used in the first case, and in the second case, borate buffer solution 0.01 M was used. A shift of the maximum light absorption of mangiferin from 365 nm to 383 nm in the borate buffer solution was observed, which may be due to the formation of a complex of mangiferin with sodium tetraborate.

Solution B with the following concentrations (mg/ml): 0.35; 0.25; 0.15; and 0.05. For this, aliquots of solution B (1; 0.7; 0.5; 0.3; and 0.1 ml) were placed in 1 ml Eppendorf tubes. 0, 0.3, 0.5, 0.7, and 0.9 ml of 70% ethyl alcohol were added. Centrifugation of the solutions was carried out for 5 min at 8000 rpm.

Capillary electrophoresis analysis was carried out at +20 kV, with capillary temperature of + 20°C, and detection was carried out spectrophotometrically at a wavelength of 383 nm, and the analysis time was 10 min. As an electrolyte, a borate buffer solution 0.01 M with a pH of 9.2 ± 0.02 was used, prepared in accordance with GOST 4919.2-2016 "Reagents and especially pure substances. Methods for the preparation of buffer solutions." Previously the capillary was washed consistently with solutions of acid hydrochloric 1 M and sodium hydroxide 1 M.

Washing between acid and alkali solutions, as well as final washing before analysis, was carried out with purified water. Washing solutions and electrolyte solutions were filtered through a Vladipor paper filter of type with a membrane diameter of 25 mm. The buffer solutions, like the test solutions, were centrifuged at 8000 rpm for 5 min. The results were processed and a calibration plot was plotted (**Figure 4**).

The preparation of purified alcohol extracts from the raw material was carried out in accordance with the procedure developed by us to determine the sum of xanthones in terms of mangiferin by UV spectrophotometry for the samples under study (the procedure described above in the UV spectrophotometry section) (**Figures 5–7**). Centrifugation of solution A was carried out for 5 min at 8000 rpm.

Based on the experimental data obtained, it can be concluded that the Caucasian penny is the largest content of mangiferin among the studied species of the genus,

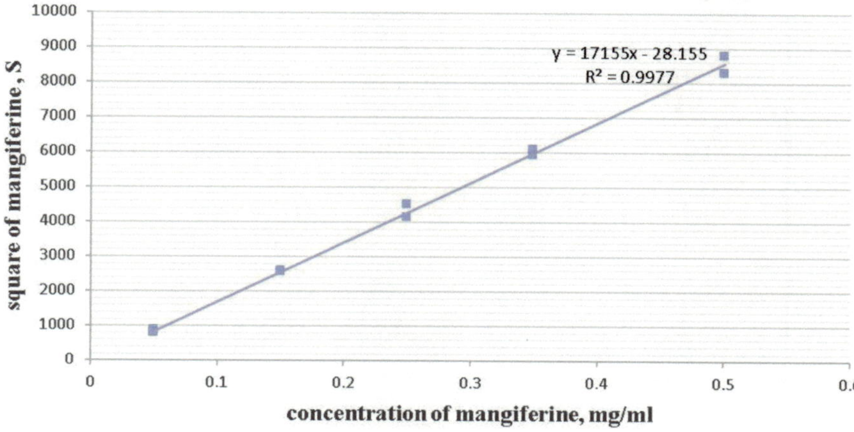

Figure 4.
Calibration graph of peak area versus mangiferin concentration in solution.

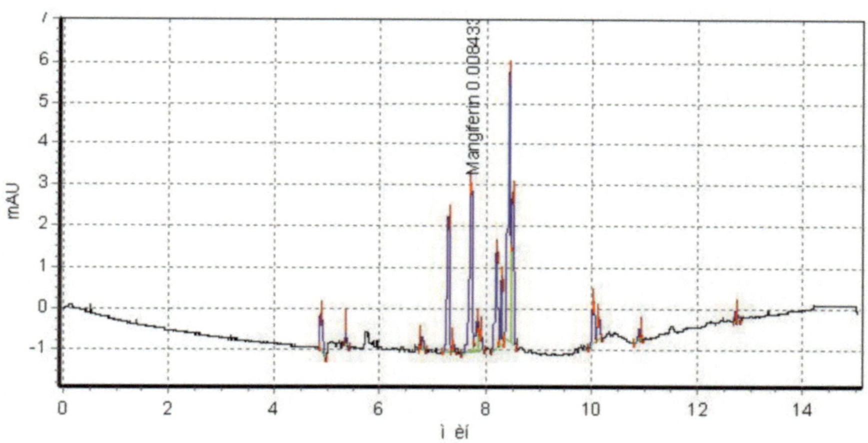

Figure 5.
Electrophoregram of extraction of Hedysarum caucasicum.

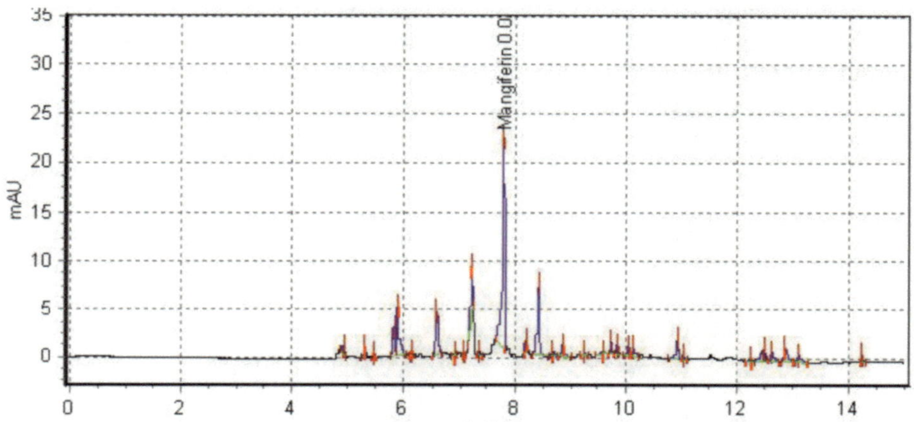

Figure 6.
Electrophoregram of extraction of Hedysarum grandiflorum *pall.*

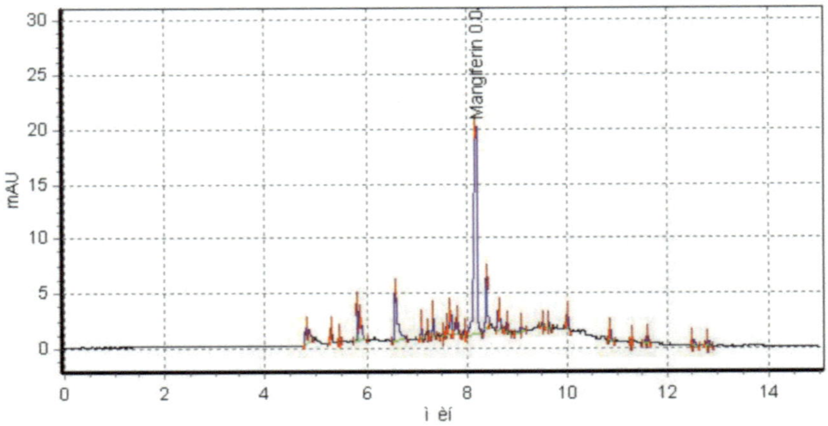

Figure 7.
Electrophoregram of extraction of Hedysarum daghestanicum *Rupr. ex Boiss.*

which confirms the assumption based on molecular genetic studies, since it is this species that belongs to the Obscura section, as well as the alpine penny used to obtain mangiferin.

12. Quantification of mangiferin in the grass of *Hedysarum caucasicum* M. Bieb. by HPLC

High-performance liquid chromatography was used to quantify mangiferin in the subjects under study. Registration of electronic spectra was carried out on a liquid chromatograph Shimadzu Prominence LC-20 AD with a degasser DGU-20A3R. A sample of mangiferin (Sigma-Aldrich, cate. M3547, USA, 2017) was used as a standard sample.

Standard mangiferin solution: A slurry of 0.0050 g (based on 100% substance) of mangiferin was placed in a measuring flask with a capacity of 25 ml, dissolved in 70% ethyl alcohol, and then the volume was adjusted with the same solvent to a mark and mixed. An aliquot of 1.0 ml of the resulting solution was placed in a measuring flask with a capacity of 10 ml, dissolved in the mobile phase, and then the volume was adjusted with the same solvent to a mark and stirred. The solution was filtered through a 0.45 μm membrane filter (**Figure 8**).

Buffer solution: In a 1000 mL measuring flask, 15.6 g of sodium phosphate of monosubstituted dihydrate was placed, which was dissolved with 200 mL of purified water and the volume adjusted to a mark, stirred. The potentiometric method determined the value of pH, which should be 4.40 $\pm$ 0.05.

Mobile phase: A mixture of buffer solution, acetonitrile, and methanol in a ratio of 81:16:3.

The developed method of quantitative determination of mangiferin in the above-ground organs of the studied type takes into account the main physicochemical properties of xanthones, and it is characterized by reproducibility, high accuracy, and allows conducting both screening assessment of various raw materials containing mangiferin derivatives and standardization of prepared medicinal plant raw materials of the *Hedysarum* L.

As a result, it was found that the largest in the above-ground organs of *Hedysarum caucasicum* found was 0.148 $\pm$ 0.003% mangiferin (**Figures 9** and **10**). The method of quantification of mangiferin by HPLC has been developed, which is

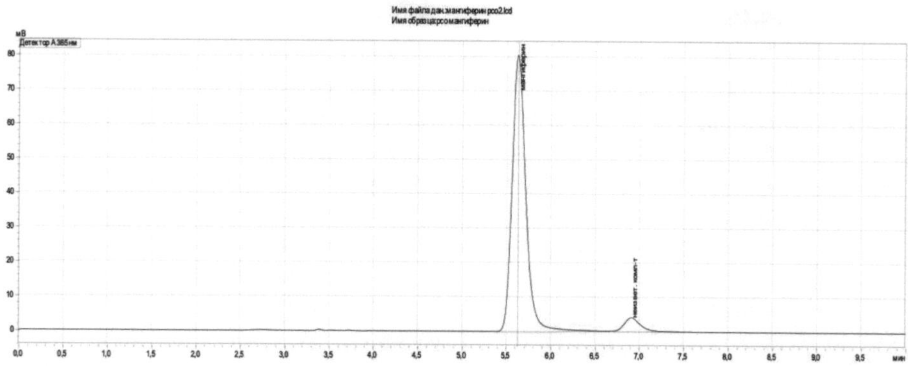

Figure 8.
Chromatogram of standard solution of mangiferin.

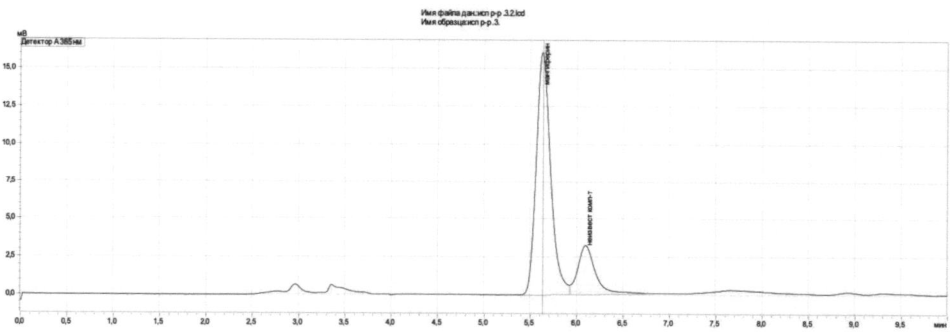

Figure 9.
Chromatogram of solution 1.

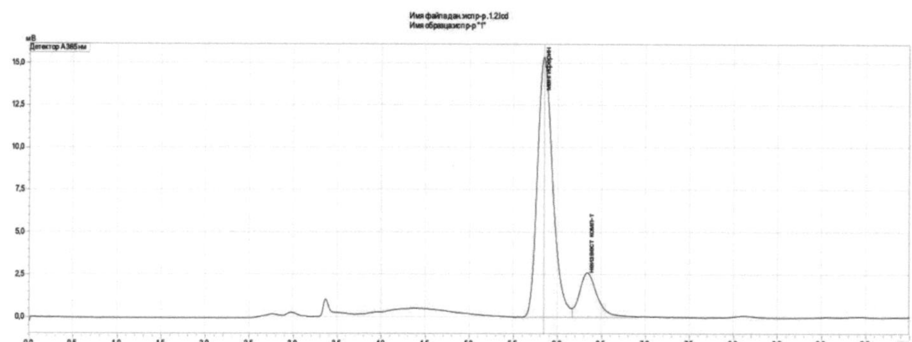

Figure 10.
Chromatogram of solution 2.

characterized by good reproducibility and tested at raw materials. The obtained results can be used in the preparation of regulatory documentation for plant raw materials. As a result, in species *Hedysarum caucasicum* M. Bieb., *Hedysarum grandiflorum* Pall., and *Hedysarum daghestanicum* Rupr. ex Boiss., biologically active substances such as polysaccharides, triterpene compounds, flavonoids, saponins, polyphenol compounds, leucoanthocyanins, and tannins were found. The quantitative content of tannins in *Hedysarum caucasicum M. Bieb*. was 5.43 ± 0.15%.

13. Amino acid analysis

Amino acid analysis in the grass of species of the genus *Hedysarum* was carried out on the basis of the FSBOU VO of the Stavropol GAU by column liquid chromatography on an amino acid analyzer according to the procedure indicated in the guest 32,195–2013. Qualitative and quantitative analysis of amino acids was carried out on an amino acid analyzer, AminoAcid Analyzer AAA 339 M (Mikrotechna). The comparative amino acid composition of the three studied samples of species of the genus growing in the North Caucasus showed that amino acids such as aspartic (1.84–2.68%) and glutamic acid (1.29–1.33%) as well as proline (0.83–1.39%), leucine (0.86–0.90%), and phenylalanine (0.58–0.63%) were found in significant quantities in the above-ground organs of the studied species. **Table 7** shows the results of a comparative analysis of the amino acid composition of the species under study.

The study of the amino acid composition during complex pharmacognostic analysis is one of the mandatory areas of research, since amino acids are involved in the biogenesis of many biologically active compounds, including xanthones. It is known that the main components of xanthone biosynthesis are acetyl-CoA,

Amino acid type		**Species**		
Types	**Structure**	**№1**	**№2**	**№3**
Monoamino dicarboxylic acids				
Asparaginic	α-Aminosuccinic acid	1.86	2.68	1.84
Glutaminic	α-Aminoglyutaroic acid	1.33	1.31	1.29
Monoamine monocarboxylic acids				
Leucine*	α-Aminoisocapronic acid	0.90	0.89	0.86
Valine*	α-Aminoisovaleric acid	0.68	0.74	0.71
Alanine	α-Aminopropionic acid	0.67	0.65	0.63
Serine	α-Amino-βoxypropionic acid	0.62	0.66	0.60
Phenylalanine*	α-Amino-β-phenylpropionic acid	0.62	0.63	0.58
Glycine	α-Aminoacetic acid	0.55	0.58	0.57
Threonine*	α-Amino-β-hydroxy-butyric acid	0.54	0.51	0.50
Thyrosine	α-Amino-β-oxy phenylpropionic acid	0.52	0.63	0.54
Isoleucine*	α-Amino-β-ethyl-β-methylphenylpropionic acid	0.48	0.58	0.54
Methionine*	α-Amino-γ-methylthion-n-butyric acid	0.04	0.11	0.10
Monoamine monocarboxylic acids				
Lysine*	α, ε-Aminocapronic acid	0.74	0.73	0.72
Arginine	α-Amino-σ-guanidine-n-valeric acid	0.55	0.56	0.56
Heterocyclic compounds				
Proline	Pyrrolidine-α-carboxylic acid	0.83	1.39	1.19
Gystidine	α-Amino-β-imidazolyl-propionic acid	0.33	0.42	0.35

**№1—Hedysarum caucasicum M. Bieb.; №2—Hedysarum grandiflorum Pall.; and №3—Hedysarum daghestanicum Rupr. ex Boiss.*

Table 7.
Comparative analysis of amino acid composition in three samples of Hedysarum L. species growing in the North Caucasus, %.

Element title	Raw material content, %	Element title	Raw material content, %
Macroelements			
K	1.47	Na	0.10
Ca	0.29	P	0.49
Mg	0.10		
Microelements			
Al	0.049	Mn	0.0025
Ba	0.0029	Cu	0.00098
B	0.00098	Ti	0.0073
Fe	0.029	Cr	0.0009
Si	0.29	Zn	0.0049

Table 8.
Elemental composition of herb Hedysarum caucasicum *M. Bieb.*

mevalonic, and shikimic acids, from which phenylalanine is further synthesized. The obtained research results can be further used in preparing a complex metabolomic evaluation of medicinal plant raw materials of *Hedysarum* species.

Spectral analysis of the above-ground organs of *Hedysarum caucasicum* M. Bieb. was carried out at the Central Testing Laboratory of Caucasian Geologic Survey JSC in Essentuki. Sample preparation of grass raw material *Hedysarum caucasicum* M. Bieb. was carried out according to the methodology of the Pharmacopea of Russian Federation. Ash was studied according to the method of the MP plant 4C atomic emission spectrometry on diffraction spectrograph.

The ash sample was evaporated into the graphite electrode cell by means of an electric arc. In equilibrium processes of excitation and reverse transition to the basic state of electrons of element atoms, emission (emission) spectra were recorded. The results are shown in **Table 8**.

Spectral analysis of herb *Hedysarum caucasicum* M. Bieb. represented by macroelements (potassium, calcium, magnesium, sodium, phosphorus) and trace elements (manganese, iron, zinc, copper, and silicon). The obtained results of elemental composition are necessary in the complex analysis of the sum of active substances of the plant.

14. Conclusion

For the first time, morphological-anatomical diagnostic signs of the species are necessary for standardization of medicinal vegetal raw materials. Indicators of caulifolar micromorphology were introduced to diagnose plant objects using the example of the genus *Hedysarum* L. Molecular genetic studies were carried out and morphometric indicators were determined, which make it possible to establish correlations between morphological, molecular genetic, and phytochemical indicators of species assigned to certain sections of the genus, as well as to predict the accumulation of xanthones in previously unearthed species.

Introduction studies of *Hedysarum causicum M. Bieb.*, *Hedysarum grandiflorum Pall.*, and *Hedysarum daghestanicum Rupr. ex Boiss.* on the territory of the Botanical Garden of the PMFI and in the Dagestan Scientific Center of the Mining Botanical Garden of the Russian Academy of Sciences, including the main phases of development, phenological spectra, were compiled.

Within the framework of phytochemical screening, BAVs such as polysaccharides, triterpene compounds, flavonoids, saponins, polyphenolic compounds, leucoanthocyanins, and tannins were found in the objects we studied. For the first time in the grass of three species of the genus *Hedysarum*, the presence of amino acids, the main part of which belongs to the group of essential amino acids, and the presence of proline and phenylalanine prove the presence of xanthones. The obtained research results can be further used in preparing a complex metabolomic evaluation of medicinal plant raw materials of *Hedysarum* species.

A comprehensive study of the qualitative analysis and quantitative content of the sum of xanthones and, in fact, mangiferin was carried out using thin-layer and paper chromatography, UV spectrophotometry, capillary zone electrophoresis, as well as high-performance liquid chromatography.

In the framework of complex pharmacognostic studies of three species of the genus *Hedysarum*, we have developed a method for quantitative determination by UV spectrophotometry of the sum of xanthones in terms of mangiferin. The technique takes into account the basic physicochemical properties of xanthones; is characterized by the reproducibility, high accuracy, and simplicity of execution; and allows conducting both a screening assessment of various raw materials containing mangiferin derivatives and standardization of the prepared vegetable raw materials.

The developed methods are tested on the above-ground organs of *Hedysarum* species, collected and dried taking into account the rules and requirements for the preparation of medicinal raw materials. As a result, it was found that the greatest quantitative content of the sum of xanthones in terms of mangiferin is distinguished by the grass *Hedysarum caucasicum* M. Bieb. ($0.62 \pm 0.021\%$). The results show the prospect of further investigation of *Hedysarum caucasicum* M. Bieb. herb as an additional source of mangiferin.

Conflict of interest

The authors declare no conflicts of interest.

Author details

Serebryanaya Fatima[1*] and Imachueva Djavgarat[2]

1 Department of Pharmacognosy, Botany and Technology of Phytopreparations of Pyatigorsk Medical Pharmaceutical Institute, Pyatigorsk, Russia

2 Daghestan State Medical University of the Ministry of Health of Russia, Mahachkala, Daghestan Republic, Russian Federation

*Address all correspondence to: f.k.serebryanaya@pmedpharm.ru

References

[1] Padmapriya K, Abhishek D, Surabhi C, Debjani D. Microwave assisted extraction of mangiferin from *Curcuma amada*. 3 Biotech. 2012;**2**: 27-30. DOI: 10.1007/s13205-011-0023-7

[2] Chriki A, Combes D, Marrakchi M. Genetic control of the conversion of dihydroflavonols into anthocyanins in flowers of *Hedysarum carnosum* Desf. Comptes Rendus de l'Académie des Sciences Series 3. 1986;**302**(16):585-588

[3] Dai R, Li K, Li Q, Bi K. Determination of mangiferin, jateorrhizine, palmatine, berberine, cinnamic acid, and cinnamaldehyde in the traditional Chinese medicinal preparation Zi-Shen pill by high-performance liquid chromatography. Journal of Chromatographic Science. 2004;**42**(4):207-210

[4] Glyzin VI, Bankovsky AI, Pavlova NS. Phenolic compounds of *Hedysarum sachalinense* and *H. brandlii* connections. Chemistry of Natural Substances. 1970;**3**:369

[5] Duan Z, Sun L, Zheng H, Yin W. Determination of inorganic elements and amino acids in the roots of *Hedysarum polybotrys* hand.-Mazz. Lanzhou Daxue Xuebao, Ziran Kexueban. 1990;**26**(2):79-82

[6] Huang Z, Cui Z, Ren Y, Zhang J, Cran M. Antiaging effect of *Hedysarum polybotrys* polysaccharide. Zhongcaoyao. 1992;**23**(9):469-473

[7] Hostettmann K, Jacot-Quillarmod A. Identification of xanthones and new arabinosides of flavone C-glucosides from *Swertia perennis* L. Helvetica Chimica Acta. 1976;**59**:1584-1591

[8] Lan Z et al. Effects of radix Hedysari polysaccharides on immunological function and transplanted tumors in mice. Zhongguo Yaoli Xuebao. 1987; **8**(3):275-277

[9] Ahangarian S, Kazempour-Osaloo S, Maassoumi AA. Molecular phylogeny of the tribe *Hedysareae* with special reference to *Onobrychis* (Fabaceae) as inferred from nrDNA ITS sequences. The Iranian Journal of Botany. 2007;**13**: 64-74

[10] Alaniya MD. Flavonoids of *Hedysarum sericeum* and *H. caucasicum*. 1983;**19**:612-613. Plenum Publishing Corporation. ХПС.-1983.-№5.-С.646

[11] Aslanukov AK, Ayrapetova AU, Serebryanaya FK. Identification and the quantitative determination of the sum of xanthones in terms of mangiferin in a grass of *Hedysarum caucasicum* Bieb. Development, research and marketing of new pharmaceutical production: Scientific papers. Pyatigorsk State Pharmaceutical Academy. 2009;(64): 11-13

[12] Beji A, Boukef K, Biard JF. Effects of artificially introduced autotetraploidy on the free phenolic acid proline in leaves of five species of the genus *Hedysarum*. II Biruniya. 1988;**4**(2): 133-144

[13] Bojnanský V, Fargašová A. Atlas of Seeds and Fruits of Central and East-European Flora: The Carpathian Mountains Region. Springer Science & Business Media; 2007. p. 1046

[14] Lin SJ, Tseng HH, Wen KC, Suen TT. Determination of gentiopicroside, mangiferin, palmatine, berberine, baicalin, wogonin and glycyrrhizin in the traditional Chinese medicinal preparation sann-joong-kuey-jian-tang by high-performance liquid chromatography. Journal of Chromatography. A. 1996;**730**(1-2): 17-23

[15] Wang T, Pan Y, Ma G, Chen L. Determination of mangiferin in qingqiliangying injection by reversed-

phase HPLC. Zhongguo Zhong Yao Za Zhi. 1997;**22**(10):602-603, 639

[16] Chennaoui H, Marghali S, Marrakchi M, Trifi-Farah N. Phylogenetic relationships in the North African genus *Hedysarum* as inferred from ITS sequences of nuclear ribosomal DNA. Genetic Resources and Crop Evolution. 2007;**54**(2):389-397. DOI: 10.1007/s10722-006-0001-9

[17] Leeprechanon W, Jutiviboonsuk DA. Quantitative determination of mangiferin isolated from leaves of *Mangifera indica* L. variety Nam doc Mai using HPTLC and its DPPH scavenging activity. In: Proceedings of the 7th International Conference of Suan Sunandha Rajabhat University; 28–29 April 2015; Bangkok, Thailand. Vol. 1(6). 2015. pp. 163-173

[18] Hai LQ, Zhang QY, Liang H, Zhao YY, Du NS. Study on chemical constituents of *Hedysarum polybotrys*. Acta Pharmacologica Sinica. 2003;**38**: 592-595

[19] Dai R, Gao J, Bi K. High-performance liquid chromatographic method for the determination and pharmacokinetic study of mangiferin in plasma of rats having taken the traditional Chinese medicinal preparation Zi-Shen pill. Journal of Chromatographic Science. 2004;**42**(2): 88-90

[20] Schieber A, Berardini N, Carle R. Identification of flavonol and xanthone glycosides from mango (*Mangifera indica* L.) peels by high-performance liquid chromatography-electrospray ionization mass spectrometry. Journal of Agricultural and Food Chemistry. 2003; **51**(17):5006-5011

[21] Jutiviboonsuk A, Sardsaengjun C. Mangiferin in leaves of three Thai mango (*Mangifera indica* L.) varieties. Indian Journal of Pharmaceutical Sciences. 2010;**6**(3):122-128

[22] Bhuvaneswari K. Isolation of mangiferin from leaves of *Mangifera indica* L. var alphonso. Asian Journal of Pharmaceutical and Clinical Research. 2013;**6**(2):173-174

[23] Kitanov G, Blinova KF. Mangiferin in some species of the *Hypericum*. Chemistry of Natural Substances. 1978; **4**:524

[24] Tibe O. Phenolics and Condensed Tannins from Sulla (*Hedysarum coronarium*) Leaves and their Biological Significance: Diss. ... Master of Science Degree in Chemistry. New Zealand: Massey University; 2003

[25] Zhang YY, Wang Q, Qi LW, Qin XY, Qin MJ. Characterization and determination of the major constituents in *Belamcandae Rhizoma* by HPLC-DAD-ESI-MS. Journal of Pharmaceutical and Biomedical Analysis. 2011;**56**(2):304-314

[26] Burlando B, Pastorino G, Salis A, Damonte G, Clericuzio M, Cornara L. The bioactivity of *Hedysarum coronarium* extracts on skinenzymes and cells correlates with phenolic content. Pharmaceutical Biology. 2017;**55**(1): 1984-1991. DOI: 10.1080/13880209.2017.1346691

[27] Rybachenko AI, Krivut BA, Georgievskii VP. Fluorodensitometric determination of a mangiferin and isomangiferin in *Hedysarum flavescens* and *H. alpinum*. Chemistry of Natural Substances. 1976;**4**:448-450

[28] Shemeryankina TB. Development of a technique of the quantitative definition of a mangiferin in raw materials by HPLC method. The state and the prospects of creation of new finished pharmaceutical products and phytochemical medicines. Kharkiv. 1990:167-168

[29] Solovyova EV, Hotsialova LI, Krivut BA, Glyzin VI, Maysuradze NI.

The maintenance of a mangiferin at types of *Hedysarum* L., grown up in Moscow areas. Plant Resources. 1983; **19**(3):356-360

[30] Wang R, Chen Y. Analysis of the constituents of *Hedysarum polybotrys*. Lanzhou Daxue Xuebao, Ziran Kexueban. 1988;**24**(3):46-50

[31] Gómez-Zaleta B, Ramírez-Silva MT, Gutiérrez A, González-Vergara E, Güizado-Rodríguez M, Rojas-Hernández A. UV/Vis, ^{1}H, and ^{13}C NMR spectroscopic studies to determine mangiferin pKa values. Spectrochimica Acta. Part A, Molecular and Biomolecular Spectroscopy. 2006; **64**(4):1002-1009

[32] Jutiviboonsuk A, Leeprechanon W. Stability of mangiferin in lotion and its antioxidant activity. Key Engineering Materials. 2019;**819**:79-84

[33] Shashi KS, Rupali MT, Saurabh KS, Chhanda CD, Satyendra KP. Antimicrobial evaluation of mangiferin and its synthesized analogues. Asian Pacific Journal of Tropical Biomedicine. 2012;**2**(2 Supplement):S884-S887

[34] Liu Y, Zhang Z, Zhang QY, Pu XP, Zhao YY. Study on chemical constituents and antioxidative activity of Radix Hedysari. Chinese Pharmaceutical Affairs. 2010;**24**: 543-549

[35] Luo F, Lv Q, Zhao Y, Hu G, Huang G, Zhang J, et al. Quantification and purification of mangiferin from Chinese mango (*Mangifera indica* L.) cultivars and its protective effect on human umbilical vein endothelial cells under H_2O_2-induced stress. International Journal of Molecular Sciences. 2012;**13**(9):11260-11274

[36] Garcia D, Leiro J, Delgado R, Sanmartin ML, Ubeira FM. *Mangifera indica* L. extract (Vimang) and mangiferin modulate mouse humoral immune responses. Phytotherapy Research. 2003;**17**(10):1182-1187

[37] Peres V, Nagem TJ. Naturally occurring pentaoxygenated, hexaoxygenated and dimeric xanthones: A literature survey. Química Nova. 1997;**20**(4):388-397

[38] Li YJ, Bi KS. Pharmacokinetics of mangiferin in rat plasma after oral administration of a single dose of suanzaoren decoction. Yao Xue Xue Bao. 2005;**40**(2):164-167

[39] Dong Y, Tang D, Zhang N, Li Y, Zhang C, Li L, et al. Phytochemicals and biological studies of plantsin genus *Hedysarum*. Chemistry Central Journal. 2013;**7**:124. journal.chemistrycentral.com/content/7/1/124

[40] Wang W, Liang H, Wang B, Tu GZ, Chen HB, Zhao YY. A new natural product from the roots of *Hedysarum multijugum*. Journal of Peking University (Health Sciences). 2005; **37**(5):532-535

[41] Wei Y, Shu P, Hong J, Qin M. Qualitative and quantitative evaluation of phenolic compounds in Iris dichotoma Pall. Phytochemical Analysis. 2011;**23**(3):197-207. DOI: 10.1002/pca.1343

[42] Yang M, Wang W, Sun JH, Zhao YY, Liang H, Guo DA. Characterization of phenolic compounds in the crude extract of *Hedysarum multijugum* by high-performance liquid chromatography with electrospray ionization tandem mass spectrometry. Rapid Communications in Mass Spectrometry. 2007;**21**:3833-3841

[43] Birdsong BA, Alston R, Turner BA. Distribution of canavanine in the family Leguminosae as related to phyletic groupings. Canadian Journal of Botany. 1960;**38**(4):499-505